S0-DRA-077

```
629.1323   A 1

Allen, John Elliston.
   Aerodynamics: the science
of air in motion.
```

PASADENA CITY COLLEGE
LIBRARY
PASADENA, CALIFORNIA

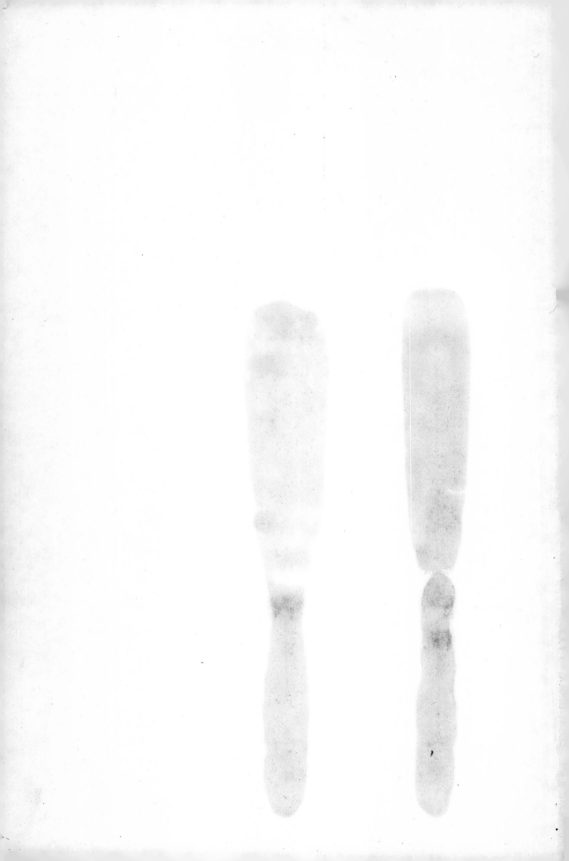

Aerodynamics
The Science of Air in Motion

Aerodynamics
The Science of Air in Motion

John E. Allen
B.Sc.(Eng.), D.N.E.C.(Hons), C.Eng., F.R.Ae.S.,
F.I.Mech.E., A.F.A.I.A.A., F.B.I.S., F.R.S.A.,
M.W.S.E.

Chief Future Projects Engineer
British Aerospace, Kingston-upon-Thames
Surrey, U.K.

McGRAW-HILL Book Company Limited
New York St Louis San Francisco

Granada Publishing Limited – Technical Books Division
Frogmore, St Albans, Herts AL2 2NF
and
36 Golden Square, London W1R 4AH

First published in the USA 1982 by McGraw-Hill Book Company,
1221 Avenue of the Americas, New York 10020

Copyright © 1982 John Allen

Library of Congress Cataloging in Publication Data

Allen, John Elliston, 1921
    Aerodynamics, The Science of Air in Motion.
    Revision of: Aerodynamics, A Space Age Survey
1963
    Bibliography: P.
    Includes index.
    1. Aerodynamics. I. Title.
    TL570.A675 1981 619.132′3 81–15560

ISBN 0–07–001074–9   AACR2

First published in Great Britain 1982 by Granada Publishing

Printed in Great Britain

All rights reserved. No part of this publication may be reproduced, stored in a retrieval system, or transmitted in any form or by any means, electronic, mechanical, photocopying, recording or otherwise, without the prior permission of the publishers.

Granada ®
Granada Publishing ®

Contents

	Preface to the First Edition	vii
	Preface to the Second Edition	ix
	Acknowledgements	xi
1	INTRODUCTION	1
2	AERODYNAMICS THROUGH THE AGES	4
3	THE NATURE OF AIR IN MOTION	8
4	AERODYNAMIC THEORY AND EXPERIMENT	30
5	NATURAL AERODYNAMICS	50
6	TRANSPORT AND INDUSTRIAL AERODYNAMICS	87
7	AERONAUTICS	118
8	AERODYNAMICS IN SPACE	161
9	AERODYNAMICS AND CIVILIZATION	186
	References	197
	Index	201

Preface to the First Edition

Aerodynamics is traditionally the subject of specialists. Its many forms and myriad applications make this inevitable, and there are, in consequence, already scores of books so entitled. This book is designed for the non-specialist, the young student, the scholar leaving school and seeking an interest for his life's work, and for the intelligent member of the public who realises that aerodynamic knowledge is both important and fascinating. The last twenty years have brought about a revolution in aerodynamics leading to our understanding the real boundaries of motion in the atmosphere of our planet. One result of this knowledge will be the flight of man-made vehicles into the very different atmospheres of Mars and Venus. This climax of earth-bound aerodynamics will initiate an extensive new field of experimental aerodynamics in other planetary 'airs'. An attempt has been made, therefore, to include a wide range of aerodynamic conditions and applications, and thus, unavoidably, some aspects are not treated completely and the subject may appear simpler than it really is. But endeavouring to cover the whole subject in a small volume is a rigorous discipline in emphasising common factors, unifying theories and mathematical concepts. Readers already familiar with basic aerodynamics may prefer to omit chapters 3 and 4, while the physical and mathematical formulae in these chapters can be passed by in a first reading without affecting the general sense. They should, however, be read later, as they contain the outline of many fundamental theories and ideas.

My aim has been to give as complete an outline as possible of the scope of aerodynamics and those physical characteristics of the air which define the types of motion commonly observed. Some methods of theory and experiment are described together with interesting applications to be found everywhere in the modern world. It is too short to be a complete textbook and there has been no attempt to concentrate on mathematical methods or expound experimental details. The book is, in a sense, an anthology, including interesting examples the author has met in his career, and it is hoped that it will form a starting point for many who are new to the subject.

Material for the book has come from many sources, some of which are given in a bibliography. My thanks go to Dr. D. J. Shapland for help and advice, to Mr. C. Scruton of the National Physical Laboratory for help on Industrial Aerodynamics, to Mr. Sidney Hoyle and Miss Jessie Lorimer for research, to Mrs. Freda M. Makin, who patiently typed from my illegible handwriting, and to Mr. R. A. Wall, who carefully read the drafts and compiled the index. The help and criticism of Kenneth and Helen Diprose during the final stages was invaluable. Chris Storey drew the diagrams. It was Tom Dalby of Hutchinson's who suggested a modern general book on aerodynamics, and Miss Frances Gibbs who, with skill and patience, helped the book along from author to printer.

Bramhall, England   J. E. A.

Preface to the Second Edition

The first edition did well in overseas editions and was translated into Dutch, Japanese, Spanish and German. Granada Television made an educational programme on it. The general structure of the book is still relevant twenty years after it was conceived, but there have been many other exciting developments.

Matters which have taken on a greater significance since 1960 include energy, pollution and noise. The greater power of calculation provided by the digital electronic computer has transformed many parts of aerodynamics and meteorology. New topics treated are aerothermochemistry, fluidics, insect flight and new classes of aircraft. Aerodynamics seems even more important to civilisation in 1980 than it did in 1960!

I am most grateful for the helpful criticism of many of my colleagues with whom I have shared several aerodynamic problems over many years. The willing advice of the following has greatly improved the quality of the text: C. L. Bore, Professor I. C. Cheeseman, Professor J. F. Clarke, R. M. Denning, Dr A. I. Fraser, I. Harris, Professor D. Howe, B. C. Kervell, Dr Ø. Ljungström, Dr E. W. E. Rogers, K. Rollins, A. C. Rudd, Professor R. S. Scorer, R. F. Spenser, Professor J. L. Stollery, Dr A. H. Wickens; and P. Wright.

My thanks also to Miss M. Bourner, Miss K. Heap and Miss S. Burge in the closing stages of preparation and particularly to Mrs Susan Clarke for the bulk of the typing.

Kingston-upon-Thames, Surrey
J. E. A.
1981

A note about units

Quantities are quoted both in Imperial and SI units wherever possible. Where metric units are used universally, e.g. in radar power and wavelength, only SI values are quoted.

Acknowledgements

N.A.S.A., Washington, USA for figs 2, 9, 70, 71; AGARD, Paris, France for figs 12, 66, 72; Department of National Defence, Defence Research Board, Ottawa, Canada for fig. 14; McGraw-Hill Book Company Inc, New York, for permission to use fig. 15 based on fig. 8.1 in Streeter's *Handbook of Fluid Dynamics* (1961); Royal Aeronautical Society for figs 18 and 74; Los Alamos Scientific Laboratory, University of California for fig. 19; National Oceanic & Atmospheric Administration, National Severe Storms Laboratory, USA for figs 20, 21, 22; Department of Electrical Engineering and Electronics, Dundee University, Scotland for fig. 23; Professor R. S. Scorer for fig. 24; National Physical Laboratory, Teddington, Middlesex for figs 25, 30, 32; National Aeronautical Establishment, Ottawa, Canada for fig. 26 (taken from Technical Paper 88 by A. G. Davenport, Division of Building Research); Nationaal Luchtvaartlaboratorium of Amsterdam, Holland for fig. 27; Dr Alistair Fraser for fig. 28; John Wiley & Sons Inc., New York for fig. 29; *Scientific American* for figs 33 and 46; General Motors Detroit, USA for figs 34 and 35; British Rail Advanced Technology Centre, Derby for fig. 36; British Hovercraft Corporation Ltd for figs 38 and 39; Swinden Laboratories, The United Steel Companies Ltd. for fig. 40; Professor T. Weis-Fogh for fig. 45; Dr Dietrich Küchemann and Pergamon Press, Oxford, UK for figs 47, 49, 62; British Aerospace Public Limited Company for figs 51, 57, 58, 59, 63; D. Napier & Son Ltd., Luton Division, The Airport, Luton, Beds., for fig. 52; American Institute of Aeronautics and Astronautics for fig. 55; Perkin Elmer Corporation, Norwalk, Connecticut, USA for figs 56 and 75; United States Air Force for fig. 61; General Dynamics Corporation, Convair Division, Fort Worth, Texas, USA for fig. 68; Jet Propulsion Laboratory, NASA, California for fig. 73; I am also grateful to M. Z. Krzywoblocki of the University of Chicago for permission to reproduce Table 5 and Sir John Mason, Director General of the Meteorological Office for Reference 14.

1
Introduction

This most excellent canopy, the air, look you, this brave o'erhanging firmament, this majestical roof fretted with golden fires

Hamlet: Shakespeare

Aerodynamics, the science of air in motion, is one of the most exciting studies of our day. The conquest of the air as it has unfolded in the 20th century has brought revolutionary changes in our knowledge and way of life. It was accomplished for the most part by flying at speeds of up to 500 mph (224 m/s) but in the last few years air speeds have risen fifty-fold, making space-flight possible and thus offering untold prospects for exploration and the advance of knowledge for centuries to come. Aerodynamics is closely linked with aeronautics which has provided its greatest stimulus, but there are other important though less publicised realms of aerodynamics concerning motion of the atmosphere itself, and many applications in industries and modern civilisation.

The air is the blanket of gases surrounding the Earth. At the surface it is a mixture of gases, mostly nitrogen and oxygen and therefore relatively dense. Like all gases, air is very mobile and is readily displaced or set in motion by solid objects or by heating. Atmospheric density is affected by pressure, temperature and humidity, and decreases with height. It is 1/10th of sea-level value at 70 000 ft (21 km), 1/100th at 160 000 ft (49 km) and 1/1 000 000th at 320 000 ft (98 km). At the very highest levels it is quite different from lower air, being highly electrified or 'ionised', and here the aurorae play. Higher still it merges imperceptibly with the very thin but measurable atmosphere of the Sun, and beyond this with the still thinner gases between the stars. Since we can now launch solid objects through all regions of the 'air' so widely defined, there is a new interest in considering these regions together, whereas formerly they were regarded as quite isolated.

The scope of aerodynamics is to observe and measure the movement of air, to evolve physical laws governing the several kinds of motion and to derive theoretical methods of predicting air motion. As this is in fact very difficult, we must continually compare theory and experiment to bring them more into alignment. Even with well-known kinds of flow such as in fans and windmills, there is still disagreement between theory and experiment. It is not certain that in time such doubts will be satisfactorily resolved, for new frontiers are being explored which, raising far greater problems, get most of the attention.

'Aerodynamics' is traditionally associated with the aeroplane, and in this context the subject deals not only with the disturbances set up in the air by moving bodies and the aerodynamic forces created on them, but with the stability and performance of flight and pressures on the surface of the aeroplane. In fact, aerodynamics is by no means restricted to the aeroplane, and in covering its various applications such topics as stability and control, which are peculiar to aeronautics, will have to be neglected. But this restriction in the treatment of aeronautics will make room for discussion of some interesting and very varied motions of air occurring in other subject fields which are not popularly associated with aerodynamics, such as meteorology (for such phenomena as winds, tornadoes and hurricanes), industrial aerodynamics (for the effects of wind on bridges, buildings and ships' superstructures), and astrophysics (for the gaseous motions amongst the stars).

Aerodynamics provides a good example of a rapidly expanding modern science. The present high rate of increase in knowledge creates problems that did not exist even fifteen years ago. Digital computers now produce prodigious quantities of data, previously beyond solution within the average lifetime. The number of professional people engaged in all scientific research and its applications doubles, approximately, every fifteen years. The sheer volume of written documents makes the task of keeping abreast of work already completed, in order to avoid costly repetition, even more difficult. We are entering an age in which the elimination of outdated or clumsy theories and facts is as important as the creation of new knowledge. This is a problem for most of us–certainly for those learning at schools and colleges, and those planning curricula and books.

This book attempts to present the whole field of aerodynamics in one volume. It is hoped that by re-arranging the explanation of aerodynamic motions in a most general way, it will be as useful to a beginner who is interested in meteorology as to one interested in aeroplanes or spaceflight. The references provided should then help the reader to follow any particular line of interest a little further. Table 1 gives an outline of the scope of aerodynamics described in this book.

There are several branches of knowledge that touch and overlap this particular grouping of 'Aerodynamics'.

Fluid dynamics is the part of theoretical physics which deals with the motion of gases, liquids, greases, oils and many combinations such as aerosols and sprays. This book selects only those parts dealing with the air in motion.

Mechanics of flight is the science of flying which includes much of the aerodynamics of this book (chapters 7 and 8) but also treats the motion and stability of the aeroplane and spacecraft acted on by aerodynamic forces.

Aeronautics can be defined as including the engineering science of flight as a whole, the aerodynamics, the flight mechanics, stability, structures, automatic systems, propulsion and engineering of vehicles used in mechanical flight.

Table 1

Main branches	Associated with	Special features
Natural (Chapter 5)	Meteorology Climates Soil and water movement Plants and vegetation	Mostly subsonic, considerably affected by heating occurring at relatively small temperature differences
(Chapter 8)	Solar prominences Aurorae	At very high altitude and near the sun electrical and magnetic effects also occur
Industrial (Chapter 6)	Air pollution Winds on structures	Mostly subsonic, wide variety of applications
Other non-aeronautical man-made	Blast furnaces and other machinery Transport vehicles	Flow in pipes is often supersonic
	Nuclear power generation	Extremely high temperature
Aeronautics (Chapter 7)	Aeroplanes and missiles Hovercraft Ballistics	Very wide speed range up to tens of thousands of miles per hour. Extreme ranges of air density and pressure. Interest lies in what the air does to bodies
In space (Chapter 8)	Spaceflight Meteorites Plasma jet	

Examples have been selected from all these branches to give a comprehensive description of atmospheric air in motion, and of its interactions with solid objects.

2
Aerodynamics through the Ages

To have gathered from the air a live tradition
Ezra Pound

Until the last two centuries, man needed very little aerodynamic knowledge. He could move, live, eat and fight without taking much notice of the air, and he became aware of its power and effects only through storms when the wind might blow away the roof of his hut, or from the draughts created for lighting fires and smelting ores. It is true that there were those gifted with imagination beyond their times, who observed the birds and wished to emulate them, but the attempts at manned flight of Icarus and Bladud were not based on any aerodynamic understanding. The ancient Greeks believed air to be one of the four basic 'elements' of which all things were constituted, the other three being earth, fire and water. Modern science has revealed quite a different table of elements, but the Greeks were right at least in assigning such importance to air, as it plays a vital part in our lives, from the very motion of air into our lungs with our first breath. Unfortunately, the Greeks' contribution to aerodynamics was small. Aristotle, for example, believed that as a body moved through the air a vacuum was formed ahead of it, which caused the motion to continue. This was long before Newton's ideas of momentum led naturally to the idea of the resistance of air to motion through it. The Greek concept was as erroneous as the phlogiston theory of burning accepted by chemists in the 18th century. Leonardo da Vinci made designs for both ornithopter (flapping wing device) and helicopter, but he was equipped neither to calculate the forces due to their motion through the air, nor to perform the necessary engineering. Some of his drawings illustrate turbulence in the air during storms.

The first theory of air resistance came from Sir Isaac Newton in 1726. He recognized that both air and water moved under similar laws and that aerodynamic forces depended on the density and velocity of the fluid and on the shape and size of a displacing object. Soon many other theoretical solutions of fluid motion problems were evolved which were restricted to flow under 'idealized' conditions, that is the air was assumed to possess constant density and to move in response to pressure and inertia. Practical interest was centred on three applications of aerodynamics: the windmill, ballistic devices (guns and cannons) and the hot air balloon. Knowledge was mostly derived by trial and error, and codes of practice did not exist. A step forward was taken during the 18th century, when experimental techniques of measurement were introduced. Benjamin Robins in England constructed a whirling arm to determine the air resistance of bodies, and a 'ballistic pendulum' to find the velocity of a bullet or shell. In the former experiment, a horizontal arm was rotated about a vertical axis by the tension of a string holding a falling weight. After a few rotations the speed of the end of the whirling arm was constant at about 25 ft/s (7.6 m/s). Test objects were mounted at the end of the arm and their air resistance altered the speed of rotation. This device was used to compare the resistance of different shapes, and to show how plate resistance changed with angle to the airflow. In the ballistic pendulum experiment, the bullet was fired into a heavy suspended block which swung through a measurable angle. The bullet speed at impact was calculated from this angle, and the mass of the block and bullet. From these tests it was learned that air resistance increases considerably as the speed of sound is approached.

During the 19th century some uncertain progress towards heavier-than-air-flight was made by gliders and powered models. In the same period the introduction of blast furnaces required large quantities of gas to be pumped efficently at high pressure and temperatures. In large bridges and buildings the dependable calculation of wind forces was needed, and with the improvement in artillery, greater precision was essential in measuring supersonic air resistance and designing bullets and shells for stable flight.

The evolution of theoretical approaches

Each part of present-day aerodynamic knowledge has a fascinating history of painstaking research, of successful and disastrous theories, and of progress more often spasmodic than smooth. Only one example will be given here, although many others have been described.[1]

The first quantitative formula for the upward lift of an inclined plate moving in the air was based on Newton's theories. This assumed elastic rebounds from the particles considered as separate 'bullets'. This theory greatly underestimates the actual truth. Rayleigh, in 1876, postulated another pattern of airflow somewhat similar to the flow set up by a plate planing on a water surface. This was a better approximation but was still not good enough. The disparity between these theories and actual measurements is clearly shown in fig. 1. Obviously, no aeroplane could be designed on this basis. It remained for the Russian, Joukowski, in 1907, to visualise the influence of the lifting plate extending by viscous effects throughout a very wide volume of the air. Formulae derived from this airflow pattern gave very close agreement with experiment. This evolution is probably the most important one in aeronautics, and took about 200 years!

The hesitating progress of theoretical and practical aerodynamics during the 19th century reflected the wide divergence between the worlds of theory

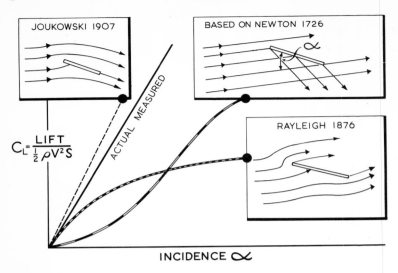

Fig. 1 Historical theories of aerodynamic lift

and experiment. Aerodynamics requires greater judgement in correlating theory and experiment than many other sciences. The Wright Brothers were the supreme example of their time of men gifted with practical skill, theoretical knowledge and insight. The aeroplane was born of this type of genius, but later progress has demanded, and will continue to demand, men of wide understanding of aerodynamic theory, mathematics and practical engineering. In many countries, special institutes have been set up to encourage this kind of mental approach and, in the USA for example, there are many outstanding aerodynamic educational centres such as the Massachusetts Institute of Technology, and the Jet Propulsion Laboratory of the California Institute of Technology.

Effects of military aeronautics

In military operations, aerodynamics has not only played its part with the aeroplane, but also in ballistics, meteorology and the flow and mixing of gases and smoke in the atmosphere. In aviation itself, although the first decade advanced slowly, the impact of military necessity caused rapid progress in the second. In the third decade, the foundations of scientific aeronautics were firmly established: large national research organizations were formed, many universities created centres of aerodynamic research, and more and more mathematicians and physicists were attracted in order to grapple with the problems of lift, drag, engine cooling, propellers and structural intricacies. There have been three major advances in aerodynamic theory, all of which emerged during the first half of the 20th century.[2] These were: *Aerofoil theory*, which extended Joukowski's hypothesis to complete aeroplane wings and propellers, *boundary layer theory*, which is the basis of understanding the nature of air resistance created near the boundary of a moving body, and *gas dynamics*, which describes the behaviour of air when compressibility and temperature changes become important, as in supersonic flight.

3
The Nature of Air in Motion

The air nimbly and sweetly recommends itself unto our gentle senses

Macbeth: Shakespeare

In order first to *describe* the motions of the air, and secondly to *calculate* them, we must map out the motions in some way. The first part of this chapter includes the visualisation of airflow and brief descriptions of the major flow 'patterns'. Methods of calculating airflows involve defining flow quantities such as speed and direction (i.e. velocity), and then working out equations which express these in terms of the geometry of the flow. Then there are physical laws which relate the pressure of the air at any point to the velocity there, and by summation these pressures give the aerodynamic force.

Visualisation

It may seem surprising that a real grasp of the nature of aerodynamic motions has only been gained in the last half century, but the invisibility of air is a peculiar hindrance to understanding its motion. We can touch a jewel, we can break a wooden strut and we can weigh a sack of flour, but although we can feel the wind and see its effects, only rarely, in nature, can we actually see it at work. Fortunately, there are now many ways of showing the motion of the air. Smoke streams can be introduced which indicate not only the flow direction, but whether it is smooth or disturbed. Shafts of light, thrown across a smoke-filled wind-tunnel, pick out the wake of a wing or body. Small tufts of wool or even fine strands of hen-feathers or cat's whiskers can show direction and oscillations. If the flow is represented in water, aluminium powder sprinkled in the liquid will indicate local motions which can be photographed with varying exposures. Alternatively, flow in water can be marked by streams of hydrogen bubbles released by electrolysis from thin charged wires arranged across it. If interest is centred on the layer of air near a surface, chemical films can be applied on which the air will create patterns by evaporation showing the direction and steadiness of the flow. In supersonic flows, the air density changes are sufficiently large to allow the air to be photographed directly, using optical systems sensitive to density changes. By these and many other means airflow can be seen, and measurements and

Fig. 2 Combusion of high energy jet fuel experiment

calculations are made from photographs (fig. 2). Flow visualisation, although a great help, can only explain some of the phenomena we want to understand. Other evidence is obtained from instruments which measure the speed and direction of the steady air current, the pressure, turbulence or oscillation, the temperature and, at high speeds, the flow of electrified particles. At extremely high speeds, very high temperatures, or very high altitudes, it is sometimes necessary to measure the magnetic field strength of the flow.

Streamlines

In a steady airflow, one can draw an imaginary smooth line which indicates the direction of flow at all points along it. This is called a streamline. Since the flow direction is parallel to a streamline, no air crosses it. This leads to the idea of a streamtube whose walls are streamlines which confine a known current of air. If the velocity increases along a streamtube, e.g. as it passes round an obstruction, the streamtube changes its cross-sectional area. If the flow becomes unsteady, the streamlines shown by the flow visualisation methods will disappear. 'Streamlining' is the shaping of bodies to promote a smooth airflow which has the low air resistance so necessary for propelled moving bodies.

Patterns of airflow

In spite of the very great freedom of movement permitted to the air it does behave according to quite distinct rules and there are several patterns of flow which occur very frequently. Some of these aerodynamic motions are illustrated in fig. 3. In this particular example the air is shown passing round

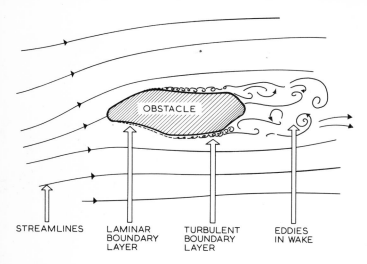

Fig. 3 Typical aerodynamic flow patterns

an obstacle which could be taken to be the wind flowing past a hill, or a projectile passing through the air. Five of these major patterns are described further in this section. They are:

  (i) a main, streamlined flow;
 (ii) a vortex or large circulatory flow;
(iii) a boundary layer with rapid changes of fluid speed near to surfaces;
(iv) a wake or ragged disturbed flow left downstream of an object;
 (v) shock waves, which occur if the flow exceeds the speed of sound.

In order to define these air patterns more clearly, it will be helpful to examine closely the motion of small regions of the air. If the flow is magnified sufficiently to indicate the behaviour of a small volume which, although small, still possesses bulk properties, this is called a fluid element. Fluid elements can have three kinds of motion:

Translation–in which elements move but do not change their direction;
Rotation–in which the elements rotate about their centres;
Distortion–which alters the shape of elements.

These elements are larger than the air molecule and it is not necessary to probe into the behaviour of individual molecules. The air is described adequately by average properties and this is called continuum flow. At very high airspeeds, however, the flow can sometimes be explained better by considering the behaviour of the molecule itself and the energy changes it undergoes.

Main flow

The simplest type of continuum flow is defined as having constant density and no particle rotation, which is a very good approximation to moving air, provided that the speed is less than the speed of sound and one does not go too close to the surface of a body. Its pattern can be shown by streamlines, and velocity (V) and pressure (p) changes are given by Bernoulli's equation:

$$p + \tfrac{1}{2}\rho V^2 = \text{constant}, H \text{ (the total head).}$$

ρ is the constant density, and $\tfrac{1}{2}\rho V^2$ is called the dynamic pressure, also referred to as q. Note that as the velocity increases, the pressure decreases, and vice versa. This is essentially what occurs with a lifting aeroplane wing. The air flowing over the curved top surface is forced to speed up, giving a reduced pressure, while the flow deflecting below is slowed somewhat, giving a higher pressure. Bernoulli's equation is important because it enables the air pressure on a body to be calculated once the air velocity near the surface is known.

In this kind of flow the air velocity can be derived from a potential function, ϕ, which has a unique value at every point. If a region of interest in a flow is mapped out in x, y and z Cartesian co-ordinates, the velocity at any point can be identified by its components u, v and w parallel to the x, y and z axes respectively. These components are given by the gradients of the velocity potential, viz:

$$u = \frac{\partial \phi}{\partial x}, \qquad v = \frac{\partial \phi}{\partial y}, \qquad w = \frac{\partial \phi}{\partial z}.$$

All points having the same value of the velocity potential function lie on an imaginary 'equipotential surface' which is everywhere at right-angles to the streamlines passing through it. A potential airfow has many points of analogy with electrostatic potential, for example, electric field strength corresponds to air velocity. Because of this, possible patterns of streamlined airflow are similar to possible patterns of electric and magnetic fields. This principle is used in the electric analogue wherein the electric field distribution is measured in an electrolyte in a tank containing a model of the body past which the airflow streams (fig. 4). In the arrangement shown, streamlines are plotted, but by rearranging the experiment with conducting walls on the other sides, lines of equal velocity potential may be drawn in.

Another property of the potential functions describing Ideal Flow patterns is that they must satisfy Laplace's equation, viz:

$$\nabla^2 \phi = \frac{\partial^2 \phi}{\partial x^2} + \frac{\partial^2 \phi}{\partial y^2} + \frac{\partial^2 \phi}{\partial z^2} = 0.$$

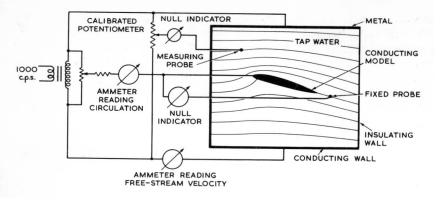

Fig. 4 Electrical analogue of a Laplace flow

This can be shown by considering the fluid flowing across a small elementary volume of air of size $\delta x \times \delta y \times \delta z$. As no fluid accumulates in such a volume, the current entering it must be the same as that leaving it. This is expressed as:

$$\frac{\partial u}{\partial x} + \frac{\partial v}{\partial y} + \frac{\partial w}{\partial z} = 0.$$

This is the Equation of Continuity. Remembering that $u = \partial \phi/\partial x$, etc., it follows that $\nabla^2 \phi = 0$.

Many functions satisfying Laplace's equation give the airflow past several shapes of practical interest, e.g. in figs 5 and 24.

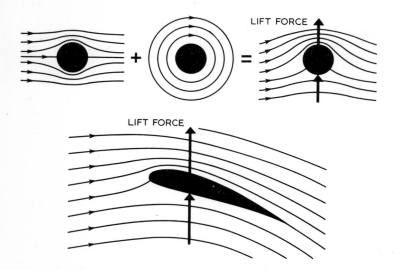

Fig. 5 Circulation and lift on an aerofoil

The streamlines in a Laplace flow round simple shapes, e.g. a cylinder, can be mathematically 'warped' to give the equivalent flow pattern round a more complex shape, e.g. a lifting aeroplane wing (fig. 5). This technique is called conformal transformation, and has proved to be a very valuable method. Another mathematical method of plotting the flow round a given object is to combine several 'sources' (points from which air is created) with several 'sinks' (points from which air is steadily removed). The position and strength of the sources and sinks is adjusted until a particular streamline has the shape of the object, the others then cover the flow field.

The calculation of flow patterns obeying Laplace's equation is a highly developed subject involving numerical methods and various methods of approximation. Such calculations start with a knowledge of conditions around the boundary of the flow. The flow field is divided into a mesh of squares of equal size, and approximate values of the stream function are first given to each intersection. If Laplace's equation is satisfied, an exact relationship should exist between values of the potential function at particular parts of each square. The errors present in the first square can then be calculated and corrections applied. Several re-calculations eventually lead to a settled set of values of the potential function. This process is also carried out by using relaxation methods on digital computers. Streamlines have been plotted automatically by electrically driven pens moving in response to the measurement of the electric potential set up across conducting paper.

If a body is symmetrical and its axis lies along the flow, streamlined 'axially-symmetric' flow can be evaluated. In practice, however, most airflows are three-dimensional with little help given by symmetry, and a good example is the aeroplane with body, wings, engine nacelles and tail. It is not yet possible to calculate mathematically airflows as complex as this, but many ingenious approximate solutions have been worked out for special cases. The 'area rule' described in chapter 7 is one example. A major difficulty is allowing for the 'interference' between the flow past one part of the object with that past another. Interference flow and forces are often only found by experiment.

Not all main flows are adequately described by potential flow. If the density varies, or heat energy is transferred, or the flow exceeds the speed of sound, the assumptions will be violated and errors will arise. Other and more complicated flows are treated in chapters 4, 7 and 8, and there is a summary in table 5.

Circulation, vortex flow, vorticity, eddies and turbulence

If an ideal flow is created round (not past) a cylinder, so that the streamlines are all concentric circles, the motion is defined as a circulation. This is measured as the integral of the velocity taken round a curve through the air enclosing the cylinder, i.e. circulation $K = 2\pi rV$ where V is the tangential velocity to the radius r. This is constant throughout the region in an ideal flow, hence the velocity is inversely proportional to the radius. If a circulation flow round a cylinder is superimposed on the flow *past* a cylinder, a lift force is produced at right-angles to the flow direction (fig. 5). This is the Magnus force which swerves spinning golf balls and is the principle underlying the Flettner rotor ship.

A *vortex* is a portion of air in such a rotational motion. A vortex may be free and persist in the air by itself, as in a tornado or in the vortex shed from an aeroplane wing. A bound vortex is one embracing a body or touching a surface. A vortex is a circulation generally quite large compared with fluid elements. (N.B. *Vorticity* is a measure of the rotational motion of a very small fluid element and equals twice its mean angular velocity.)

Eddies are another kind of rotational flow larger than 'elemental' vorticity but usually not so definite or constant as a vortex. They are to be found, for example, in the leeward windflow round buildings, and their presence is shown by leaves and dust.

Turbulence, which is another time-varying quantity in an airflow, is the irregular velocity fluctuation superimposed on the mean flow. It can be thought of as composed of extremely variable eddies of a wide range of size and direction, but it can be treated on a statistical basis. Its intensity is measured as a ratio of the velocity increments to the mean speed. It is commonly observed as the general gustiness of the wind.

The boundary layer

The air in motion divides itself very neatly into the main flow, where viscosity or fluid friction plays a negligible part, and the 'boundary layer', confined to a region very close to a surface, which is predominantly influenced by viscosity.

Air particles very close to a solid surface encounter molecular forces and adhere to it so that the air speed at the surface is zero. The air speed increases rapidly away from this adhered layer (thus creating viscous forces) until the main flow is reached. We are interested in the thickness of the boundary layer which may be only $\frac{1}{10}$ to 1 in (2.5 to 25 mm) thick over an aeroplane wing, or scores of feet over a hilly countryside, and how this grows along the surface, and how rapidly the airspeed increases as we travel across it. The shearing action near the surface creates skin friction drag, which is especially important in aeronautics. At supersonic speeds large amounts of heat are generated in boundary layers amounting to several kilowatts per square foot.

(i) *Laminar boundary layer*. The change of air speed from the surface to the top of the boundary layer is nearly linear (fig. 6). The air moves as if it were in very thin layers, each sliding over its neighbours. Differences in speed between successive layers introduce molecular viscous forces but no large vortices are produced.

When an airflow first encounters a solid surface, the boundary layer is often laminar. Further downstream the layer thickens as more and more air is slowed down by the surface friction. Usually the laminar flow cannot be maintained in a thick boundary layer and at the transition point a more violent eddying motion sets in to give a turbulent boundary layer.

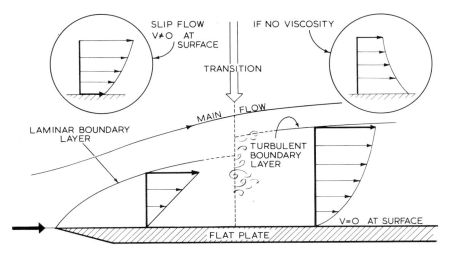

Fig. 6   Boundary layer flow

(ii) *The transition point* is extremely sensitive to several factors such as surface shape, roughness, steadiness of the oncoming airflow, temperature difference between surface and air and many other subtle effects. The uncertainty in locating the transition point particularly in model experiments has led to many erroneous aerodynamic laws and incorrect deductions.

(iii) *Turbulent boundary layer*. It is not difficult to visualise this for it can be seen in water alongside ships in motion, but it is less easy to evaluate. Because extra energy is absorbed in starting up the rotating turbulence of the molecules, the skin friction is greater (3 to 10 times as much), and the layer thickens more rapidly along the body surface than in laminar flow. Heating effects are also much greater. Some approximate quantities of boundary layers, and how they change with different flight conditions, are shown in Table 2, which has been calculated for a smooth, flat plate.

Table 2

Regime	Boundary layer height		Frictional stress at surface	
	(in)	(mm)	(lb/ft^2)	(Pa)
Laminar (subsonic)	1/10	2.5	1/20	2.39
Turbulent (subsonic)	1	25	2/10	9.58
Turbulent (supersonic)	1	25	1	47.9
Laminar (hypersonic)	1/10	2.5	1½	71.8

The explanation of the boundary layer, first put forward by Prandtl in Germany in the 1920's, represented a tremendous step forward in theoretical aerodynamics. What is so remarkable is that the whole idea of viscosity taking effect in a restricted region rather than in some uniform way throughout the main flow was quite unexpected.

Wakes

The disturbed flow downstream of an object is the wake. If the body is streamlined and the boundary layer is laminar the wake will be thin, containing only the small vorticity of the boundary layer, and will not take away much energy with it. Rough bodies, bluff shapes or lifting bodies exposed at a high angle of incidence to an airstream have large wakes, consisting of a confused flow composed of large eddies, turbulence and violent changes of general pattern. Such wakes absorb energy and create high retarding or 'drag' forces on the object. Examples of wakes are given on pages 16 and 167.

There are wakes behind railway carriages, in the lee of mountains and buildings, beyond the crest of ocean waves subject to wind, behind satellites orbiting the earth, and behind chimney stacks. Because of the variety of object shapes and airflow, wakes take on a very great variety of patterns. Because of the random nature of the flow, wakes are not easily calculated, except in special cases. There is one type of wake, however, which has a pattern amenable to measurement and theory, and is called the vortex street (fig. 7).

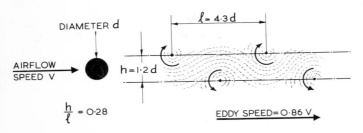

Fig. 7 Von Kármán vortex street

The wake which forms behind many objects is responsible for creating oscillating lateral forces on the body. This can produce 'singing' notes in wires, violent destructive forces in bridges, and other troubles. The eddying wake created by a jet impinging on a sharp edge produces high frequency vibrations in the air, which is the basis of organ pipes and the wood-wind instruments of an orchestra.

Shock waves

The basic patterns of airflow so far described apply broadly to any speed of flow, but there are other qualitative flow changes which arise when the air velocity approaches, or exceeds, the speed of sound. Why is the speed of sound apparently so important? All disturbances to air initially at rest are brought about by a series of pressure changes such as, for example, a small explosion generated at a point, or by the wing of an aeroplane cleaving its way through the air, parting it into the flow beneath and the flow above. Such a disturbance moving the air molecules travels at a definite speed through the air which is the speed of sound, provided that the disturbance is not too violent. The speed of sound has a definite value.

$$a = \sqrt{\gamma RT}$$

where $\gamma = \dfrac{C_p}{C_v} = \dfrac{\text{specific heat at constant pressure}}{\text{specific heat at constant volume}}$,

R = gas constant, and T = absolute temperature (K).
For air this is:

$$a = 65.8\sqrt{T} \text{ ft/s} \quad \text{or} \quad 20.1\sqrt{T} \text{ m/s}.$$

which is 1116 ft/s (340.3 m/s) at normal temperature and pressure.

This is the speed, too, at which the sound of thunder travels, i.e. it takes roughly 5 seconds to travel 1 mile (3 s per km). If the pressure disturbance moves through the air at less than the speed of sound, the outward radiating pressure waves will extend well ahead of the disturbance. At supersonic speed this cannot happen, and disturbances are only propagated sideways and behind the body. The region ahead of these disturbances has been called the zone of forbidden signals and it has a definite boundary. For a simple point disturbance or one created by a sharp body, the region is a cone with its apex at the front of the disturbance and its semi-angle is $\sin^{-1}(1/M)$ (fig. 8). M is the Mach number.

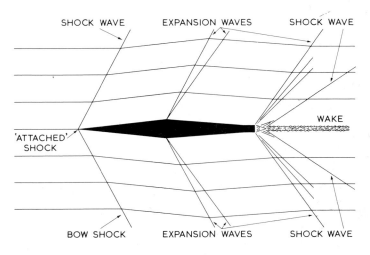

Fig. 8 Shock and expansion waves in supersonic flow

The boundary line is called a 'shock wave'. A shock wave is remarkably thin, usually only a few molecular free paths thick, i.e. about 10^{-5} to 10^{-7} in (2.5×10^{-5} to 2.5×10^{-7} cm). It arises because when air is compressed its temperature rises, and since the velocity of sound increases with temperature strong waves travel faster than weaker ones and overtake them. Thus several pressure disturbances created over a large flow region behind a shock wave will 'pile up' at the shock wave. 'Wave' used in this sense is analogous to the rollers breaking on a shelving sea beach and is not like the sinusoidal swell in the open sea. The reason for the breakers on the shore is also an accumulation of waves of varying velocity and results from the decreasing depth of water.

As a shock wave passes a point, the pressure, density and temperature rise and there is transfer of energy to the air. If a shock wave passes over another surface it generally thickens the boundary layer and changes the air pressure aft of this point. An expansion wave occurs when a supersonic flow turns through a negative angle, i.e. in the opposite sense to one creating a shock wave (fig. 8). A complex shape such as an aeroplane flying supersonically is surrounded by a very complex set of shock and expansion waves which change with Mach number and the inclination of the aeroplane to the direction of flight. The X-15 (fig. 9) is a good example of this.

Fig. 9 Photograph shows shock waves of a $4\frac{3}{4}$ in (12 cm) model of the X-15 research aeroplane flying at Mach 2.5

We have now looked at basic flow patterns and their different characteristics. There are several means by which air is set in motion, as shown in table 3.

Table 3

Air flow created by	Examples		Chapter
Moving bodies	Aeroplanes, missiles		7
	Bullets, meteorites		8
	Ships, cars, bicycles		6
	Men, birds and insects		7
Addition of mech- anical energy†	Blowing by mouth Propellers, fans Pistons in internal combustion engines.	} Engines	6
Addition of heat energy†	Turbojets, ramjets, rockets Internal combustion engines.	} Engines	7
	Cooling devices–'radiators' Heat exchangers Exhausts and chimneys	} Waste Heat	6
	Convection flow due to differential heating of earth or sea. Intense fires.	} In nature	5
Explosions‡	Volcanoes Nuclear and thermonuclear explosions Gun efflux		3
Inertia forces	Movement of large atmospheric masses over planetary surfaces arising from combined effects of gravity and planetary rotation.		5, 8
Magnetohydro- dynamic forces	Plasma jet Electric rockets Pinch effects Cosmic electrodynamics		8

† A distinction is made between natural and man-made examples because the latter are usually in pursuit of some manipulation of energy in an engine and the aerodynamic processes are controlled in special ways.

‡ Explosions could be thought of as examples of sudden addition of heat energy leading rapidly to pressure changes.

The resulting flow patterns in these examples usually combine several of the simpler basic flows already described, but the aerodynamic problems are quite distinct because of differences in airspeed, the geometry of the restraining boundaries and the nature of the energy changes. Some of these energy transfer processes are described in the next section.

Aerodynamic forces and other transfers of energy

In any airflow, there are important changes of energy that occur throughout the air itself, and between the flow and solid surfaces. There are three main classes of energy transfer: mechanical forces (pressures), heat and electromagnetic forces.

Aerodynamic pressures and forces

Interest in the pressure of the air on a small region of a solid surface arises for four reasons:

- (i) such information is directly relevant to the lifting force of an aeroplane wing;
- (ii) the surface must be strong enough to withstand the pressure without collapsing or distorting;
- (iii) a definite amount of air may have to be removed from the outside flow either to measure its characteristics, to ventilate a building, or to feed a fan or an engine;
- (iv) gas may be ejected from that point, e.g. an exhaust or a petrol jettison vent.

The surface pressure results from the speed and density of the main flow, and from the speed changes and energy losses which have occurred while the flow has been passing over the surface up to the point in question. It will thus depend on the state of the boundary layer and the angle of the surface to the main flow. If there is a large intake at the point, as in a jet engine, this inward flow will appreciably change the airflow round the body. Local roughness or bending of plates on the surface will also considerably modify the pressures. Pressures can be steady or may vary with time. In the latter case, the oscillations can be large and slow, such as an eddying flow, or small and rapid as in the high frequency sound vibrations created by jet engines or rocket exhausts.

A pressure coefficient is defined thus:

$$C_p = \frac{\text{pressure}}{\text{dynamic pressure}} = \frac{P}{\tfrac{1}{2}\rho V^2}.$$

Figure 10 shows the pressure at various points on a roof in a wind. The C_p values are roughly independent of the wind speed, so that at 50 mph (22.4 m/s) the peak suction at point A would be 3.8 lb/ft² (182 Pa) and at 100 mph (44.7 m/s), 15.3 lb/ft² (733 Pa). If the wind direction changes, the flow alters and the C_p values will alter too.

Pressure distributions can be drawn for aeroplane wings, space vehicles, rockets, motor cars, etc. The total aerodynamic force on a body or surface is formed by adding together the pressures, allowing for variations in magnitude and direction. Forces are conventionally measured as in fig. 10. The resultant force is divided (resolved) into a drag force parallel to the main airflow, and a lift force perpendicular to it. The corresponding non-dimensional coefficients are the drag coefficient and the lift coefficient.

$$C_D = \text{drag force}/\tfrac{1}{2}\rho V^2 S \qquad C_L = \text{lift force}/\tfrac{1}{2}\rho V^2 S$$

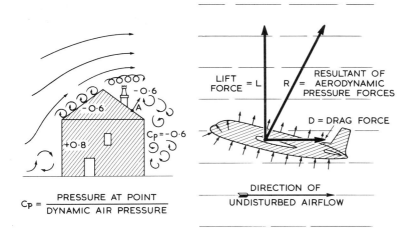

Fig. 10 Aerodynamic pressures and forces

where S is a representative surface area of the body. (The significance of this formula is explained on pages 33 and 34.)

Heating, free and forced convection

In addition to the obvious kinetic energy content of an airflow, heat transfer is very important. At very high speeds, particularly, the airflow is violently disturbed by obstacles and the energy is transformed into heat. The temperature of a flow rises whenever it is slowed down, and also increases where chemical combustion occurs. Heat energy passes to or from surfaces whenever the temperature of the air differs from that of the surface. This is a very extensive part of aerodynamic motion which we meet in several different forms, e.g. supersonic flight, meteorology and engines. As in any heat transfer process, heat can pass from a solid to any adjacent gas by conduction, convection and radiation. Transference by the first process is small, and by the last is dominant when the surfaces get extremely hot as in a jet exhaust or a meteorite. Convection is the most frequent mechanism of heat transfer in natural aerodynamics and in most cooling processes found in industry and machinery. If the main flow is sufficiently slow the convection can itself create an aerodynamic motion. In *free convection* the air is heated by the warmer surface, its density decreases and is displaced by the cooler and denser surrounding air so that it carries off heat from the surface. In *forced convection*, heat is transferred to an airstream passing relatively rapidly over the surface by a combined viscous convective process. In chapter 4 the physical quantities that enter into these motions are described and, in chapters 5 and 6, some applications are given.

Electromagnetic forces

If an airflow is at a sufficiently high temperature and low pressure, it becomes ionised and therefore electrically conducting. In such a flow a magnetic field will set up electromotive forces which can be significantly large or even exceed the mechanical and viscous forces. The electrical and magnetic effects interact with the inertia, viscosity and heat transfer quantities, and very involved flows result which require complex physical and mathematical descriptions. Such phenomena are included in the subject of magnetohydrodynamics (M.H.D.) discussed in chapters 4 and 8.

Complications

Moving air carrying solid or liquid particles[3]

The interaction between very small particles moving in an airflow will depend principally on viscous, inertia and electric charge forces, and occasionally on temperature buoyancy effects. The number of geometrical and physical combinations between air, particles (solids and liquids), surfaces and applications is so great that brief, general description is impossible. Important cases include dust deposition, filtering, froths, radioactive fallout, diffusing of salt spray, wood smoke (fish curing), and impact of rain particles on fast-moving bodies. The impact of two liquid droplets shown in fig. 11 involves a complicated type of airflow.

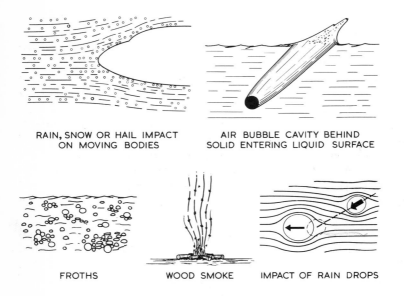

Fig. 11 Airflow with solid and liquid particles

Air entrainment is the air motion caused by a fast-moving liquid (surface or jet) which induces a motion in the surrounding air by viscous interaction at the boundary.

Explosions

An explosion leads to a rapid expansion of air, usually from a relatively small central source. The initial flow is supersonic and shock waves travel rapidly away from the source. As the volume of air affected grows so the pressure diminishes, and it is often necessary to know just how much the pressure will be at a particular distance from the explosion. This is in general a rather complex calculation but some indication is given by the simple relation applying to an intense blast radiating in three dimensions. The total energy of the explosion is E; R is the radius of the spherical wave; S is its radial speed and its pressure rise is Δp after a time t.

$$R = \left(\frac{E}{\rho}\right)^{1/5} t^{2/5}, \qquad S = \frac{2}{5}\left(\frac{E}{\rho}\right)^{1/2} R^{-3/2}, \qquad \Delta p \propto \frac{1}{R^3}$$

where ρ is the initial air density. Such relationships are used in measuring the strength of nuclear explosions at a distance. Explosion waves interact with earth surfaces and obstacles causing complex secondary waves. Behind a strong explosion shock follows a sharp negative pressure. The 'sonic' bang of supersonic aeroplanes is a special case of an explosion wave, as is also the cracking of a whip.

Aerodynamic noise

The propagation of small pressure disturbances in air was introduced on page 16 where it was stated that they travel at the speed of sound. In this section the subject is expanded and the distinction is drawn between the wide ranging subject of acoustics and those sounds that are generated as a result of an aerodynamic disturbance, e.g. the noise of a jet engine exhaust. Acoustics is the science which treats the creation and transmission of sound by any means. It deals with the physics of sound which includes reflection, refraction, diffraction, interference and absorption of sound waves not only through the air but in liquids and solids too. Since sound is perceived in the ear acoustics includes also the behaviour of this organ and how air vibrations are conducted to the eardrum. Ultrasonics is a special kind of high frequency sound that can probe into materials, without damaging them, in seeking flaws and also in medical diagnostics. This is indeed an enormous subject in its own right and whilst recognising the classical matters within the science of acoustics (Lord Rayleigh, 1896†) we shall here concentrate on those kinds of sound (or noise) which result from aerodynamic flows, i.e. not those produced by vibration of

† Lord Rayleigh (1896) *The Theory of Sound*. 2nd Edition. Macmillan.

solids as in diesel engines, railway trains or machinery. The flows of interest are particularly the noise of jet engines, propellers, rotors and rockets, shock waves and sonic booms and vortices, boundary layers and turbulence. This subject has been dubbed aeroacoustics by Goldstein[4] and sound generated aerodynamically by Lighthill.[5]

We begin by defining the quantities which are conventionally employed to measure sound levels, then proceed to describe some of the physical aeroacoustic processes and how they are approximated to by various theories. Comparisons with experiment and the means used to reduce noise complete this brief survey, but there are references elsewhere to specific applications, e.g. hovercraft, trains, aircraft and spacecraft propulsion.

Sound (of aerodynamic origin) is rarely of a single frequency. The energy is distributed over many frequencies, but the total effect can be measured. *Sound pressure level* is measured in decibels† and is:

$$20 \times \log_{10} \frac{\text{Sound pressure at a point}}{\text{A reference pressure}}$$

The reference pressure for air is usually $2 \times 10^{-4} \mu$bar. *Sound pressure spectrum level* refers to a specific frequency within a band whose width is 1 Hz. These definitions should not be confused with the power or energy which is defined as

$$\textit{Sound power ratio} = 10 \times \log_{10} \frac{\text{Sound power per unit area}}{\text{Reference power per unit area}}$$

This is measured also in decibels and is sometimes referred to as sound intensity level, specific sound-energy flux level or sound energy flux density level. Although these are indices of magnitude they do not help to explain the nature or origin of the sound.

Noise has two primary effects; firstly to create damaging oscillating pressure at close range, e.g. the hot high velocity exhausts of reheated jet engines or rockets can destroy structures very quickly by acoustic (vibration) damage. Secondly there is the noise nuisance inflicted on communities at much greater distances. The latter has proved to be of considerable complexity and many alternative criteria have been developed over the last fifty years, and particularly since the advent of jet aircraft. In rare cases the vibration effect of sound can modify an aerodynamic flow; reference[5] instances the modification of a fishtail gas burner flame in response to certain notes on the 'cello.

A modification to the simple sound pressure level definition is the A-weighting introduced in the 1930s. This weights the sound amplitudes at different frequencies in accordance with a person's hearing sensitivity and then takes the sum. Sound so measured is expressed in dB(A) units. It has been observed that background noisiness varies considerably, e.g. a camp site

---

† $p/p_0 = 2$ represents 6 dB; $p/p_0 = 3$ is 10 dB; $p/p_0 = 10$ is 20 dB; p/p_0 of 1000 = 60 dB; p/p_0 of 16 = $4 \times 4 = 12$ dB + 12 dB = 24 dB.

in the Grand Canyon has been measured to be 16 dB(A) and a house alongside a busy highway would be nearer 75 dB(A). Duration and time of day also affect community acceptance of disturbing noise which is accounted for by other criteria, e.g. Composite Noise Rating (CNR), Perceived Noise Level (PNL), Noise Exposure Forecast (NEF), Noise and Number Index (NNI) or even Psychological Assessment of Aircraft Noise Index (PAANI). The 'noise footprint' beneath an aircraft is a contour of constant noise level measured in PNdB or dB(A). This would be influenced both by the intrinsic noise and also by the rate at which the aircraft climbed away.

Aerodynamic sound from aerospacecraft originates from many different sources. The shock wave with very local but intense pressure differences is registered on the ground by its signature, i.e. pressure variation with time. The noise from jet exhaust arises both from the intense turbulence at the boundary where it mixes with the slower moving external air and also from shock waves embedded in the jet. Fans, turbines and compressors have another kind of rotational noise of rather similar origin to that of a siren. Combustion noise also escapes through the gas turbine and can be detected outside the aircraft. The boundary layer disturbances are a source of noise as are also the main wing tip vortices and those shed from the edges of flaps and drag brakes. When landing with wheels and undercarriage down and various open doors and exposed cavities there is a noticeable addition to the sound from their eddies, interferences and turbulence.

There are two fundamental problems in a theoretical threatment of aeroacoustics and myriads of complicating circumstances, depending on individual noise sources. The first step is to identify the noise creating sources throughout the aerodynamic field and these will differ for all those kinds just described. The second is then to apply acoustic theory to add up the effects of the aerodynamic sources distributed throughout the volume of the aircraft and the external flow disturbed by it and then relate these to the summation of noise perceived at particular points some distance away from the aircraft.

Some theories employ a simple elementary source which assumes that sound is radiated from a mathematically small point by an impulsive transfer of mass or momentum. The strength of the sound wave and its position after a given time can be determined.

Three different types of source are almost certainly all present in the flow fields generated by an aircraft and its engines.

The Monopole is the creation and annihilation of mass at a point and is a highly efficient sound generator relative to the systems energy. It is characterised by both a fluctuating mass and fluctuating external applied force. Examples of such sources are to be found in:

(a) the vibration of a structural panel;
(b) unsteady heat production in a combustor leading to flow expansion and contraction.

The Dipole is the close juxtaposition of two equal and opposite monopoles which almost cancel each other out as sound energy sources and as a result is highly inefficient as a sound generator while the energy which escapes is highly directional. Such a system is characterised essentially by zero mass change and a fluctuating applied external force. Examples of such sources are to be found in:

(a) rotating turbomachinery with interactions between the pressure fields of adjacent blade rows;
(b) the generation of unsteady flow around a body in a boundary layer and in flow separations.

The Quadrupole is the close juxtaposition of two equal and opposite dipoles which cancel the requirement for an externally applied force. As a result it is even less efficient than a dipole as a sound generator which is just as well since its generating capacity varies as an 8th power of the characteristic velocity.

This type of source requires neither mass change nor an externally applied force.

The main example of such a source is to be found in the turbulent mixing shear layer region created by a uniform jet into a uniform free air field.

These are applied typically as shown in fig. 12. It will be noted that the sound power depends on which kind of pole is used, which depends on the nature of the sound source.

Source type	Source associated with:	Source acoustic representation	Sound–power relationship	Examples
Monopole	Pulsating flow	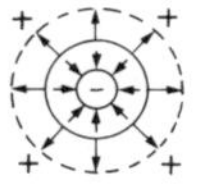	$\rho L^2 V^4/a_o$ ($\rho A V^3 M$)	Droplet combustion Pulse jets Inlet and exhausts of reciprocating machines
Dipole	Unsteady flows close to surfaces	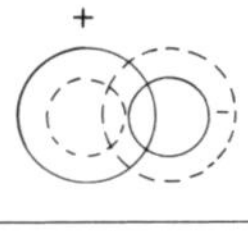	$\rho L^2 V^6/a_o^3$ ($\rho A V^3 M^3$)	Fan blade noise Boundary layer noise
Quadrupole	Free mixing of exhaust flows into atmosphere	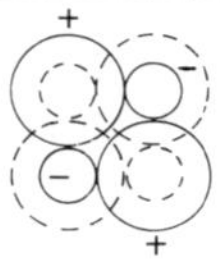	$\rho L^2 V^8/a_o^3$ ($\rho A V^3 M^5$)	Jet noise Valve noise

Fig. 12 Aeroacoustics

The intricacy of the mathematical relationships which result from the application of such fundamental concepts can be illustrated by three typical equations of aeroacoustics:

$$\frac{\overline{D}^2\rho'}{Dt^2} - \frac{\partial^2 p'}{\partial x_i^2} - \frac{\overline{D}\rho'}{Dt}\frac{\partial \bar{c}_i}{\partial x_i} - \bar{\rho}c'_j\frac{\partial^2 \bar{c}_i}{\partial x_i \partial x_j} - \left(2\bar{\rho}\frac{\partial c'_j}{\partial x_i} + \bar{c}_j\frac{\partial \rho'}{\partial x_i} + \rho'\frac{\partial \bar{c}_j}{\partial x_i}\right)\frac{\partial \bar{c}_i}{\partial x_j} = 0$$

which represents acoustical sound propagation in non-uniform flow at rest.

$$\frac{\partial^2 \rho}{\partial t^2} - a_0^2 \frac{\partial^2 \rho}{\partial x_i^2} = \frac{\partial^2}{\partial x_i \partial x_j}(\rho c_i c_j - \tau_{ij}) - \frac{\partial}{\partial x_i}(f_i + mc_i) + \frac{\partial m}{\partial t} + \frac{\partial^2}{\partial x_i^2}(p - a_0^2 \rho)$$

which is from Lighthill's inhomogeneous wave equation, and

$$\frac{D^2 r}{Dt^2} - \frac{\partial}{\partial x_i}\left(a^2 \frac{\partial r}{\partial x_i}\right) = \gamma \frac{\partial c_j}{\partial x_i}\frac{\partial c_i}{\partial x_j} + \frac{D}{Dt}\left(\frac{\gamma}{c_p}\frac{Ds}{Dt}\right) - \frac{\partial}{\partial x_i}\left(\frac{\gamma}{\rho}\frac{\partial \iota_{ij}}{\partial x_j}\right)$$
$$+ \frac{D}{Dt}\left(\frac{\gamma}{\rho}m\right) - \frac{\partial}{\partial x_i}\left(\frac{\gamma}{\rho}f_i\right)$$

which is Phillips' converted wave equation which takes account of the aircraft noise source moving through the atmosphere, i.e. not at rest.

One concludes from the foregoing that aerodynamic sound not only has many constituents creating noise by several different mechanisms but is extremely difficult to calculate in a given case or predict for a new class of aircraft wing, design or engine. Great strides have been made in understanding the nature of aerodynamic sound by these theoretical treatments which are of course complemented by thorough experimental measurements both in full scale, wind tunnels and acoustic laboratories. Wind tunnels must be specially designed or modified by means of mufflers, acoustic linings and turbulence screens to suppress, as far as possible, the pressure fluctuations of the tunnel air itself, excited by the driving fans. An indication of experimental difficulties encountered is that the noise radiated by a jet is about 1% of the energy of the turbulence which creates it and the turbulence energy is only 1% of the kinetic energy of the jet flow. All this work lies behind some of the means now being employed to reduce noise which will next be described.

The noise of existing jet engines is lowered by 'hush-kits' usually in the form of suppressor nozzles fitted over the exhaust so that the single circular jet is subdivided into a large number of smaller separate flows sometimes circular but often into radial lobes. These operate by changing frequency spectra and reducing effective radiating surface area. New jet engines, designed since noise reduction became a major objective, are quietened by acoustic duct liners fitted to internal surfaces over which pass both inlet and exhaust airflow. These are typically perforated metallic surfaces backed by several small air cavities which are tuned to absorb air vibrations of harmful frequencies. The siren noise of the front fan is almost cured by the elimination of the fixed inlet guide vanes originally fitted to swirl the air before meeting the first set of compressor blades. The 'siren' effects inevitably associated with subsequent stages can be reduced by varying the number of blades between them so that resonances or 'beats' do not develop. Mufflers are additional streamwise surfaces that split the flow into smaller sections. However, the most significant factor in reducing jet engine exhaust noise resulted from a major change of design to improve propulsion efficiency. The first jet airliners used existing military gas turbines which were 'pure jets' having small diameter high velocity exhausts. To improve propulsive efficiency steps were taken to lower the exhaust velocity and increase the size of the exhaust outflow to reduce kinetic losses in the wake. First came the by-pass engine with an unburnt compressed airflow surrounding the inner hot exhaust, and then the fanjet, characterised for example on the Lockheed Tristar where most of the thrust is provided by the cool, slower wake from the ducted fan which completely surrounds the much smaller, but still hot, core flow exhaust. The overall effect of these changes has given remarkable improvement over two and a half decades, viz. table 4.

Table 4

In service date	Engine	EPNdB (at 1750 ft/533 m) sideline	Aircraft
1956	Avon, pure jet	108	Caravelle VIR
1963	Spey, bypass jet	102	BAC 1-11 500
1972	RB 211, turbofan	90	L–1011/RB 211-22B
1981	New technology	92	B757/RB 211-535
1985	Advanced technology	80	

By mounting two jet engines on a wing so that one is above and one is immediately below it reduces overall noise by 3 dB since the lower jet shields the upper one so that some downwardly radiating noise is either reflected or absorbed when it encounters the lower jet. Sometimes tailplanes and fuselage surfaces can be designed to shield engine exhausts. Quite new designs of jet engine have been researched for new supersonic transports. New methods of mixing hot and cold flows from the engine exhausts and the use of moving mechanical suppressors are expected to reduce engine noise by 15 dB.

The aircraft designer's contribution to noise reduction includes the positioning of engines so that their exhausts and/or inlets are shielded from the ground by wings, tails or fins, closing doors around undercarriage wheel cavities after they are lowered and minimising the number of gaps between slots and flaps. These contributions are termed 'aircraft self-noise' which become of more importance as jet engines become quieter. Even the vibration of skin panels is receiving attention. The overall effect of all these engine and airframe improvements is to reduce the noise 'footprint' around an airport from 36 square miles (93.2 km^2) of the Boeing 707 of 1965 to the 5 square miles (13 km^2) of the Airbus A 300 of 1974. The sound level at the boundary of the footprint was taken to be 90 PNdB in this example.

Severe legislation has been introduced by national and international aviation authorities of which the most famous is the US F.A.R. Part 36. This laid down permissible noise levels related to aircraft weight to be attained at specified dates in the future. New aircraft such as the Airbus A 300 were able to respond to this and in fact is better than required. Others were retrospectively modified and even older ones sold off. The test conditions to demonstrate compliance with such requirements are complex as befits any aspect of aerodynamic noise: aircraft power, speed and height must be measured; also atmospheric conditions like temperature gradient, and rain; measurements should be made over generally flat terrain. In the 1970s the new emphasis on fuel economy, which could be improved to some extent if noise regulations were relaxed, is being seen however as an additional requirement; the great progress in quietening the increasing volume of air traffic will also continue.

4
Aerodynamic Theory and Experiment

And weave but nets to catch the wind
Webster

This chapter links the qualitative descriptions of flow patterns in chapter 3 with the applications described in chapters 5, 6, 7 and 8. We cannot proceed very far into these varied types of aerodynamics solely on the basis of ideal flow. There is a surprising number of physical properties of the air which affect the way it flows and which have to be measured and allowed for.[6]

The physical quantities influencing aerodynamic motion

Three fundamental properties of the air are its temperature, density and pressure.

Temperature (T) is proportional to the average kinetic energy of the air molecules. The higher the temperature, the greater the molecular motion. At the very high temperatures associated with high air-speeds energy can also be absorbed internally by the molecules inducing them to rotate and vibrate. When this occurs the air can no longer be considered as an 'ideal gas' and we speak of 'real gas effects' to denote the state where internal energy changes have also to be taken into account.

Density (ρ) is the mass per unit volume. At sea level, is 0.002378 †slugs/ft^3 (1.225 kg/m^3) at normal temperature (15°C). ρ decreases with altitude and its ratio to sea-level density (ρ_0) is the relative density $\sigma = \rho/\rho_0$.

Pressure (p) is the force per unit area exerted on or by the air. The sea-level pressure is 14.696 lb/in^2 (101.3 kPa).

These three quantities are related in the ideal gas equation $p/\rho = R^*T/w$ where R^* is the universal gas constant, and w is the molecular weight of air.

Gravity (g) also enters into several phenomena, particularly those involving heating and buoyancy, and in meteorology. It is the acceleration which would

---

† Aeronautical people use the slug as a unit of mass which equals 32.2 lb. If a mass of 1 slug is acted upon by a force of 1 lb weight it will accelerate at 1 ft/s/s.

be experienced by an unrestrained mass near the earth's surface arising from the pull of the earth's gravity. Its sea-level value is 32.2 ft/s^2 (981 m/s^2) approximately.

At high speeds the air becomes compressed when encountering an object, and other quantities are involved which can be derived from the gas equation:

Compressibility (K) represents the proportional decrease in volume resulting from the application of external pressure on a mass of air.

Coefficient of thermal expansion (β) occurs in convection and buoyancy effects.

Specific heat (C) is the quantity of heat needed to raise the temperature of a unit mass of air by a unit temperature difference. It has two values: C_p when the heat is added at constant pressure, and C_v when heat is added at constant volume. C_p exceeds C_v and the ratio $C_p/C_v = \gamma$ frequently occurs in calculations especially when heat changes in air are brought about by compressibility. It enters directly into the velocity of sound, $a = \sqrt{\gamma RT}$. R† for air is 3090 ft lb/slug/°C (287 J/kg K).

Three other properties of moving air are related to the mechanisms by which momentum, energy and mass are transported on a molecular scale. They are respectively viscosity, thermal conductivity and diffusivity.

Viscosity (μ) is the measure of the resistance of air to a shearing deformation. It is most simply demonstrated in a laminar boundary layer where adjacent parallel layers of air have different speeds. The viscosity is the force per unit area needed to maintain a unit velocity difference between layers a unit distance apart. The viscosity of air increases with temperature but is virtually independent of pressure. $\mu = 3.73 \times 10^{-7}$ slug/ft/s (1.78×10^{-5} Pa s).

[*Kinematic viscosity* (ν) $= \mu/\rho$ is thus named because its units include those of length and time only. ν varies with temperature and pressure.]

Thermal conductivity (λ or k) is the quantity of heat that flows through a unit of area per unit of temperature difference. For air at 15°C it is 4.07×10^{-6} CHU/ft s/°C (0.0255 W/mK).

The *diffusion coefficient* (D_{12}) expresses the rate of diffusion of one gas (1) through another (2). Such a mixing occurs with water vapour in sweating, or in transpiration cooling when hydrogen is passed through a porous surface in contact with a hot boundary layer to produce a temperature relief.

The major transport properties may be summarised by the three simple laws and coefficients, viz:

For viscosity: τ (shear stress) $= \mu \dfrac{du}{dy}$ (Newton's law)

For thermal conductivity: q (heat flux) $= \lambda \dfrac{dT}{dy}$ (Fourier's law)

---

† Not to be confused with Reynolds number.

For diffusion: j (mass flux) $= D_{12} \dfrac{dC_1}{dy}$ (Fick's law)

where C_1 is the concentration of gas species (1); D_{12} refers to the mixing gases. u is fluid velocity at distance y from a surface.

Thermal diffusivity (α) is related to λ and is a measure of the transfer of heat by diffusion analogous to viscous motion. It is equal to $\lambda/\rho C_p$.

In magnetohydrodynamic flows, additional physical quantities are involved, for instance:

Magnetic permeability (μ) measured in henrys/m,
Charge density (ρ_e) coulombs/m^3,
Electrical conductivity (σ) mhos/ft or siemens/m.

The preceding quantities relate to the air itself. There are others which describe the condition of surfaces in contact with an airstream especially in relation to the friction and heat transfer in a boundary layer.

Emissivity (ε) expresses the heat radiated from a hot gas or surface in the Stefan-Boltzmann law $q = \varepsilon b T^4$. The constant b = 2.78×10^{-12} CHU/ft^2 s/°C^4 (5.67×10^{-8} W/m^2/K^4). The emissivity measures the ability of a body to lose heat by radiation referring to a 'black body' as a standard for which $\varepsilon = 1$. At orbital re-entry speeds the heat radiated from a surface can exceed that conducted to the surface.

Surface roughness enters into many aerodynamic phenomena, particularly the boundary layer drag. No simple specification of roughness size exists but it can usually be divided into a 'small grain' component and a wave component. For the former, generalised data is available which is applicable to a wide range of roughness as shown in fig. 13. Roughness values can be ascribed to different types of metallic surfaces in pipes or on aeroplanes and also to terrain such as mud-flats, sea, dust, snow, or grass.

Molecular properties of surfaces. At very high altitudes and speeds, when the airflow may be highly dissociated and ionised, complex interaction processes occur between a surface and the impinging air particles. Molecules may rebound in much the same way as a rubber ball (specular reflection), or they may be absorbed at the surface and another molecule emitted from elsewhere (diffuse reflection). Surfaces usually show more diffuse than specular reflection. In the case of ions or atoms, their rate of recombination will depend on the catalytic efficiency of the surface. These affect the drag, the temperature and the quality of reflected light.

The theoretical dilemma

To solve even simple aerodynamic problems theoretically, and to include all the preceding physical quantities, is quite impossible mathematically. In practice, however, the majority are negligible, if not everywhere in the flow field, at least in regions of the flow which are reasonably well isolated from each other, and each having special features. The boundary layer is one such region predominantly affected by viscous effects. The main flow is another, provided that it is drawn round the outside of the boundary layer taking its thickness into account. Within each flow region, equations can be established but these are not always easily evaluated. Many theoretical solutions are now available for particular types of flow, which work tolerably well provided they are used in appropriate situations. Figure 15 shows some different regimes of aerodynamic flow each characterised by a major effect. The significance of Mach number (M) and Reynolds number (R) will be explained later, but for the moment should be taken as aerodynamic scales which help to separate out the influences of compressibility and viscosity. Eleven regimes are shown in fig. 13 and also their relevance to different applications. For example, at very low R where inertia is negligible, aerodynamic forces vary as the velocity. This happens with the airflow in mine shafts and in micrometeorology. When inertia forces predominate, (high R), as in aeronautics, aerodynamic force is proportional to (velocity)^2.

Dimensional analysis

The quantitative relationships between the flow properties in any aerodynamic zone can be formulated by the method of dimensional analysis. It is convenient to express all the aerodynamic quantities in terms of the three basic physical quantities length (L) mass (M) and time (T). The convention is to write, for example, $[V] = LT^{-1}$ and [force] = [Mass × acceleration] = MLT^{-2} with similar relations for pressure, density and viscosity, etc. The principle of dimensional homogeneity states that in a physical equation the dimensions must be consistent throughout. How this helps to establish with more certainty an aerodynamic relationship will be shown by one important example, viz: the aerodynamic force (F) on a body immersed in a moving, viscous, compressible gas. Let the size of the body, inclined at an angle α to the flow direction, be defined by a typical length, l, the air having a speed, V, density, ρ, viscosity, μ, and sound velocity, a. On the plausible assumption that the force will depend in some way on these quantities, we may write that:

$$F = f(\rho, V, l, \mu, a, \alpha) \tag{1}$$

Dimensional homogeneity, requiring that the dimensions of mass, length and time must be consistent throughout, provides three equations leaving three unknowns still to be determined.

The result is that:

$$F = \rho V^2 l^2 f_1\left(\frac{Vl\rho}{\mu}\right) f_2\left(\frac{V}{a}\right) f_3(\alpha) \tag{2}$$

where f_1, f_2, f_3 are unknown functions. This equation is more precise than the original one as it gives a direct relationship between F, ρ, V and l leaving as still undetermined the three functions, which it will be noted are dimensionless. These dimensionless ratios have a special significance in theoretical and experimental aerodynamics as they help to determine the aerodynamic regime of a particular flow.

In the equation above, $Vl\rho/\mu$ is the Reynolds number R, which is the ratio of inertia to viscous forces. V/a is the Mach number M, an index of the compressibility effects, and α is the incidence which expresses the commonsense idea that the airflow and hence aerodynamic force on a body depends on its angle to the flow. The equation is now

$$F = \rho V^2 l^2 f_1(R) f_2(M) f_3(\alpha) \qquad (3)$$

Similar expressions can be derived for many other quantities such as heat transfer, vortex shedding, gravity waves, etc. Some of the more important non-dimensional ratios are:

Knudsen number $Kn = x_m/L$, where x_m is the mean free path length of an air molecule (the average distance between molecular collisions or encounters), and L is a typical length of a body. This is relevant at very high altitudes where the air behaves similarly to Newton's theory (page 6).

Froude number $F = V^2/gl$, the ratio of inertia force to the gravity force on a mass of air. This is met in buoyancy problems as, for example, when a mass of warmed air moves upwards through a gravity field. In aeronautics it is negligible, but can be important where liquid droplets of water, etc. are entrained in the flow.

There is a large group of non-dimensional numbers dealing with heat and mass transfer. In the following, q is a quantity of heat, l, L or D are typical lengths.

Stanton number $St = -q/\rho V C_p \Delta T$, is a measure of the heat flow through a surface where $\Delta T = T_r - T_w$, $T_r =$ recovery temperature $= f(M, Pr)$ and $T_w =$ wall temperature.

Prandtl number $Pr = \mu C_p/\lambda = \nu/\alpha$, is a property of air only, and is the ratio of momentum diffusivity to thermal diffusivity.

Péclet number $Pe = Vl/\lambda$, where λ is the thermal conductivity, applies particularly to heat transfer from a body to air at very low velocity. (Note that $Pe = R \times Pr$, for small temperature differences).

Grashof number $Gr = \rho^2 \beta g l^3 \Delta T/\mu^2$, arises e.g. in the case of free convection flow through a narrow annular slit between circular cylinders. In a vertical wall the height is the characteristic length.

Nusselt number $Nu = -qD/\lambda \Delta T$, is important in free and forced convection as a measure of heat flux. $Nu = St \times R \times Pr$ for forced convection and $Nu = St \times Gr \times Pr$ in free convection.

Schmidt number $Sc = \mu/\rho D_{12}$ expresses the ratio of viscous and mass diffusivity.

Lewis number $Le = \rho C_p D_{12}/\lambda = Pr/Sc$, is important in the heating of re-entry vehicles.

An example of non-dimensional coefficients in aerodynamics

The drag force acting along the boundary of a flow is called the skin friction and, since it arises from the viscous forces, it would be expected to depend on the Reynolds number. Figure 13 shows the skin friction on a flat plate aligned

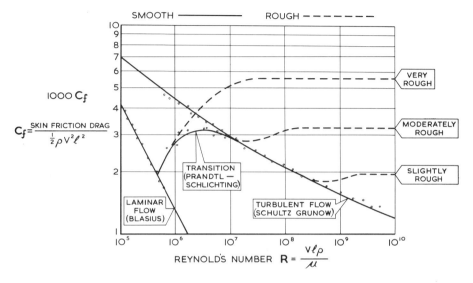

Fig. 13 Boundary layer skin friction–Reynold's number

with a uniform flow. A consequence of the dimensional method is that the skin friction coefficient $C_f = \text{Force}/\frac{1}{2}\rho V^2 l^2$ is related to the Reynolds number only, which takes into account size, velocity, density and viscosity, in one parameter. For smooth surfaces there are theoretical relationships which agree well with experimental results when the boundary layer flow is laminar or turbulent. Roughness also has a profound effect at least as important as the variations with Reynolds number.

Experimental modelling for different aerodynamic regimes

Turning now to equation (3), dimensional analysis gives the general result:

$$\begin{pmatrix}\text{An aerodynamic}\\ \text{quantity of inter-}\\ \text{est}\end{pmatrix} = \begin{pmatrix}\text{Product of powers of}\\ \text{several physical flow}\\ \text{quantities}\end{pmatrix} \times \begin{pmatrix}\text{Unknown functions}\\ \text{of non-dimensional}\\ \text{ratios}\end{pmatrix} \quad (4)$$

When the unknown functions have been found, the expression above is complete. This is done by experiment, in which the airflow is artificially arranged to permit measurement of the lefthand side and the first part of the right-hand side of equation (3), whence the desired functions are obtained by division. Experimental skill comes in designing the experiment to eliminate conflicting effects of more than one function of the non-dimensional numbers.

It is clearly difficult to make these experiments on full-sized aircraft (or buildings), and moreover this information is required for designing the aircraft before it is built. Because of the non-dimensional character of the relationships, however, realistic experiments can be made on small models, provided that the non-dimensional numbers, e.g. R, M, Pr, etc., are made the same as in the full scale. Under these restrictions the unknown functions of the aerodynamic parameters will be the same as in full scale, and the full-scale quantities can then be evaluated from equation (4) or the example equation (3).

There are many kinds of aerodynamic experiment depending on the quantity to be measured and the aerodynamic regime. In laboratory experiments, the air may be at rest and the test body may be moved (as in a ballistic pendulum), or air may be blown past a fixed model (as in a wind-tunnel). Some phenomena can be demonstrated in water, others are best measured in free flight through the air either on models with radio telemetry or with full-scale aircraft. In fig. 15 the various aerodynamic regimes are classified

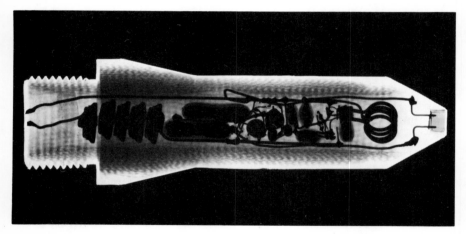

Fig. 14 X-ray photograph of telemetry transmitter in small ballistic model

either by the type of flow, e.g. viscous, compressible, free molecule, etc., or by speed, i.e. subsonic, supersonic, hypersonic.

Aerodynamic regimes

The regimes are summarised in table 5.

In the final part of this chapter, a few special features of some of these regimes will be described, together with experimental techniques.

Table 5

	Regime	Possible Mach no. and other characteristics	Chapter
Meteorological fluid dynamics	(Hydrodynamics)		5
Incompressible fluid–continuum flow	(Hydrodynamics)	Constant density	5, 6
Subsonic: pseudo-continuum, or physics of gases	(Gas dynamics)	$M < 1$	5, 6, 7
Compressible medium: gas	(Gas dynamics)	Variable density	6, 7
Transonic: pseudo-continuum, or physics of gases	(Gas dynamics)	$M \cong 1$	7
Supersonic: pseudo-continuum, or physics of gases, slip flow	(Gas dynamics)	$1 < M < 7$	7
Aerothermochemistry: combustion, re-entry	(Gas dynamics)	Variable chemistry all M	7, 8
Hypersonics, hyper-ballistics (rarefied gas); slip flow	(Gas dynamics)	$\sim 7 < M \sim 15-20$	7, 8
Free molecule flow, Newtonian gas dynamics (rarefied gas)		$\sim 15-20 < M$	8
Fluid dynamics involving electromagnetic phenomena		Ionosphere, radiation	7, 8
Meteoritical fluid dynamics (meteors)		$100 \leq M$	8
Fluid dynamics including nuclear reactions		Hydrodynamics of the Sun	8
Celestial fluid dynamics		Star clouds, plasma	8

Subsonic flow *(up to M = 1)*

The main flow outside the boundary layer is well represented by the ideal streamlined flow described on page 11. Bodies designed for subsonic flow are streamlined, having rounded front ends and thin or pointed rear ends. The density is virtually constant up to speeds of 300 mph (130 m/s) and the simple Bernoulli's law of $p_0 = p + \frac{1}{2}\rho V^2$ applies. The exact expression includes other terms depending on Mach number, viz: $p_0 = p + \frac{1}{2}\rho V^2 (1 + \frac{1}{4}M^2 + \frac{1}{40}M^4 + \ldots)$ and hence, as M increases, compressibility begins to have an effect, increasing $(p_0 - p)/\frac{1}{2}\rho V^2$ by 6% at M = 0.5, and 22% at M = 0.9. At high

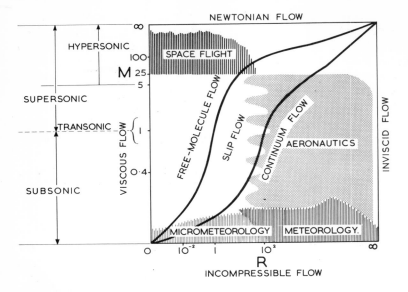

Fig. 15 Aerodynamic regimes and applications

subsonic speeds $M > 0.75$, say, compressibility alters the flow pattern and shock waves occur when the local speed reaches the speed of sound as, for example, above an aerofoil at the point of maximum thickness. This Mach number, for a given shape, is termed M_{crit}, the critical Mach number. Up to this speed, the pressure and lift coefficients at a fixed incidence increase; a relationship frequently used for this trend is the Prandtl–Glauert rule, i.e. a function of $1/\sqrt{1 - M^2}$.

Most of the first experiments made were on models driven through the air by such means as the whirling arm, but a great step forward was taken by the introduction of the wind-tunnel in 1871, in which a reasonably smooth airflow created by a fan was blown over a fixed model. The convenience of being able to measure aerodynamic forces on a balance and to change the angular setting of the model to the airstream was offset by the problem of straightening out the airflow in the wake of the fan and reducing the resulting small-scale turbulence as much as possible. The evolution of wind-tunnels, their design, and the interpretation of experimental results represents a vast and important part of aerodynamic science. A typical wind-tunnel is shown in fig. 16. Various corrections must be made to the measurements for the imperfections of the apparatus, which include (i) the extra drag of wires and struts that support the model, (ii) the constraining effect of the finite size of the airstream in the tunnel that can falsify lift, drag and moments, (iii) the non-representative position of boundary layer transition due to low Reynolds number, (iv) absence of air intakes or jets, etc. and (v) the absence of heat additions to the air as in an air-cooled engine or 'radiator'.

At subsonic speeds, Reynolds number is of primary importance. There are a few full-scale tunnels for complete aircraft with working engines at air speeds up to 250 mph (110 m/s). Aircraft models are usually made of hard wood at scales of ½ down to scales of about 1/30th. (The scales for buildings, chimneys and bridges must be much smaller, i.e. less than $\frac{1}{1000}$.) Reynolds number ($R = Vl\rho/\mu$) cannot always be made correct in a wind-tunnel as the speed (V) increase necessary to balance the reduction in size (l) would approach the speed of sound and introduce undesired compressibility effects. The disadvantage can be overcome by increasing the density of the air (ρ) in the tunnel by pumping it to a high pressure.

At high subsonic speeds models are often made in steel to withstand the air loads and heat and give accurate, smooth surfaces.

Transonic flow (*from* $M = 0.75$ *to* $M = 1.2$ *approximately*)

Transonic flow has no hard and fast limits but could be said to apply near $M = 1$ and when both subsonic and supersonic flows occur. However, even in supersonic flight (i.e. at $M > 1.2$), a blunt leading edge will have a locally subsonic flow over it, behind the shock wave, and flow in the boundary layer will be subsonic too. At exactly $M = 1$ plane shock waves occur which travel outwards perpendicular to the direction of motion of the body. This condition needs a special approximate theory of airflow, as many of the sub- or supersonic theories are not calculable at $M = 1$. The theory can also be used to some extent within the range of Mach number 0.85 to 1.15, but the success of this depends on the shape of the body and the aerodynamic quantity in question. Drag coefficient reaches a maximum value near $M = 1$ and the interference drag between bodies, wings and tails of aeroplanes can be very great. This can be appreciably reduced if the aeroplane is designed in accordance with the 'area rule' which states that the combined cross-sectional area of a complete aeroplane should vary smoothly along its length. The aeroplane shown in fig. 68 has been designed on this principle.

A subsonic wind-tunnel cannot be used for transonic speeds as the laterally displaced shock waves are reflected off the walls and produce spurious pressures and forces on the model. This was a great hindrance to experiments at these speeds and instead free flight models were either launched from high-flying aircraft or propelled by rockets from the ground. Eventually, special wind-tunnels were built having walls ventilated by slots or perforations which absorb shock waves instead of reflecting them. Both free flight and tunnel methods are now in use.

Supersonic flow

In a supersonic flow ($M > 1$), the air velocity of the main flow everywhere exceeds the speed of sound, and shock and expansion waves occur wherever a solid boundary changes its inclination to the airflow (fig. 8). Compressibility effects are now of primary importance and variations in the density of the air have to be taken into account. The problems are profoundly influenced by the generation of heat through shock waves and viscous effects in the boundary layer. At low supersonic speeds the temperature rise resulting from the compression through a shock wave perpendicular to the flow direction is given by:

$$\frac{T_2}{T_1} = \left(\frac{7M_1^2 - 1}{6}\right)\frac{5 + M_1^2}{6M_1^2},$$

where T_1, M_1 apply before shock, and T_2 afterwards. The pressure rise across a normal shock is also related to the change in Mach number, viz:

$$\frac{p_2}{p_1} = \frac{1}{6}(7M_1^2 - 1), \quad \text{actually} \quad \frac{1}{\gamma + 1}[2\gamma M_1^2 - (\gamma - 1)]$$

since $\gamma = C_p/C_v = 1.4$, i.e. 7/5 (see page 31). If the shock wave is inclined to the flow, other relations apply.

There is a point at the front of a moving body called the stagnation point where the air slows down to come to rest at the surface. In so doing it becomes heated, and the stagnation temperature T_0 is given by the formula

$$\frac{T_0}{T_1} = 1 + \frac{M^2}{5} \quad \text{or} \quad \left(1 + \frac{\gamma - 1}{2}M^2\right).$$

T_1 is the absolute temperature of the ambient air flowing past the body, i.e. about 220K in the stratosphere. T_0/T_1 is 1.2 at M =1, 1.8 at M = 2 and 4.2 at M = 4, so the stagnation temperature rises sharply at supersonic speeds.

Large supersonic wind-tunnels need many thousand horsepower to drive them and large heat exchangers to remove surplus heat from the airstream without disturbing the flow distribution (fig. 16). The construction of large steel supersonic tunnels poses several problems: the surfaces must be smooth to ensure uniform flow, they must expand thermally without distortion and the shape must be changed for each Mach number. This is done either by alternative sections of tunnel or flexible walls set to an exact predetermined shape by multiple automated jacks. The models are also of steel and are mounted on rear sting supports to minimise drag interference.

Aerothermochemistry

The subject of aerothermodynamics combines the two sciences of aerodynamics and thermodynamics and hence includes supersonics and hypersonics. Aerothermochemistry[7] describes another kind of airflow in which chemical reactions occur with exchange of chemical energy and flow energy (by which is meant kinetic energy of mass motion plus thermal energy) as in a combustion process or during atmospheric re-entry; in the former, where a fuel is burnt in air, chemical energy is converted into flow energy; in the latter various chemical changes that absorb energy from the flow are brought about between the constituents of the air itself by the intense shock waves and

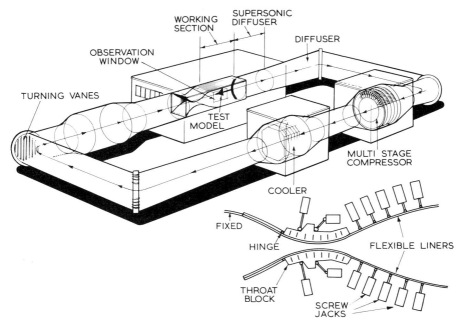

Fig. 16 Supersonic closed circuit wind-tunnel

friction at hypersonic speed. It is convenient to describe here the essential relationships of aerothermochemistry with applications on pages 168 and 174. Unlike continuum flow, which deals with average fluid properties, aerothermochemistry has to take account of the energy changes which occur at the molecular level. Gas molecules can absorb and exchange energy with their neighbours, with whom they collide, in a very large number of ways and hence this subject takes on a complexity far greater than those aerodynamic theories so far described. The time scale for changes in the internal-energy state of a polyatomic molecule or the chemical composition of a chemically reacting mixture is the average interval between these intermolecular encounters. The time taken to accomplish the energy change is referred to as the relaxation time (τ) and varies from a few molecular collisions for translational or rotational change to many thousands of encounters for a chemical reaction or vibrational energy exchanges. In any given aerodynamic situation a typical time of flow can be defined, e.g. the time taken for an average molecule to pass through a combustion chamber in a jet engine (related to its length and speed of the airflow)–τ_{flow} say. From the chemistry of the reactions there will be another time corresponding to the completion of the particular process, e.g. mixing, combustion or diffusion which is the relaxation time τ. The non-dimensional ratio of these two times, viz. τ_{flow}/τ is known as the Damköhler number D. Now if D is small the relaxation time is long compared to the flow time and the flow is said to be 'frozen' since the relaxation effects can to a first approximation be ignored. Conversely, with D large the reaction time is short, and the effect changes relatively very actively; such flows are termed equilibrium or near-equilibrium flows. As the nature of the chemical process, say, changes throughout the volume of an airflow the value of D can alter dramatically. To give a humble example: the Bunsen burner mixes air with hydrocarbon gas passing up the tube prior to combustion into a weak deflagration flame. In the cold gas stream within the tube, D is very small and the mixture is essentially chemically frozen. Heat conduction forward from the hot downstream regions of the flame into the oncoming flow raises the rate of chemical activity dramatically; D for a given gas particle increases to a level that makes possible a very rapid approach to chemical equilibrium, and the consequent liberation of chemical energy continues to heat the incoming fresh reactants and so sustains the flame/flow system.

An unexpected application of the methods of aerothermochemistry is the airflow of dusty gases. Temperature changes and the drag of individual (assumed) spherical particles allow an equivalent diffusion to be specified leading on to the description of acoustic and explosion waves. The occasional spontaneous combustion of wheat dust carried in the air in grain silos is an important application of this work.

Numerical computation of the complicated combustion-chamber flow in real gas turbine engines has reached a high state of thoroughness using powerful calculation methods available with large digital computers. Gas turbine air flow is continuous (unlike that in an internal combustion piston engine) but the actual chemical processes to be modelled are most complex, e.g. the reaction mechanisms of kerosene burning in air have to account for at least sixteen different reaction steps. Even the basic conservation equation to allow for several chemical species (in this case J) becomes formidable:

$$\rho \left\{ v \frac{\partial}{\partial r} m_J + \frac{w}{\partial \theta} \frac{\partial}{\partial \theta} m_J + u \frac{\partial}{\partial x} m_J \right\} = R_J$$
$$+ \frac{1}{r} \frac{\partial}{\partial r} \left(r \Gamma_J \frac{\partial m_J}{\partial r} \right) + \frac{1}{r} \frac{\partial}{\partial \theta} \left(\Gamma_J \frac{\partial m_J}{r \partial \theta} \right) + \frac{\partial}{\partial x} \left(\Gamma_J \frac{\partial m_J}{\partial x} \right)$$

Heat addition, radiative effects, droplet mixing, evaporation and soot formation can all be handled by the model and agreement with experiment has proved to be highly accurate. The matter cannot rest however, and more experimentation to derive better models is proceeding in many countries. Combustion in air at hypersonic speed is treated on page 159. The aerothermochemistry involved with spacecraft re-entry is discussed on page 165 et seq.

In this highly specialised aerodynamic field the dynamics of relaxing gases has emerged as a special study in its own right as a result of a unique collaboration between the molecular physicist and the aerodynamicist.

Hypersonic flow and real gas effects

Hypersonic flow is usually taken to apply at Mach numbers exceeding 5. At these speeds the bow shock wave from a body lies back close to the downstream surface (fig. 69). A special theory of air motion has been evolved for this case (Van Dyke). At the high temperatures created in a hypersonic flow, the assumption of a perfect gas is no longer accurate. Figure 17 shows

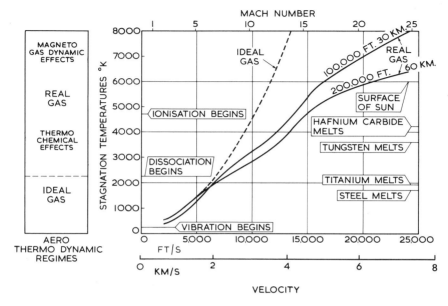

Fig. 17 High speed flight and temperature scale

how the stagnation temperature increases with Mach number. The supersonic law greatly exaggerates the temperature above M = 5 as the air molecules at high temperature absorb heat in vibration (internal motion of atoms), dissociation (molecules dividing into atoms), electronic excitation (electrons absorbing energy) and ionisation (free electrons liberated in the flow). Their heat content is no longer proportional to temperature, and the specific heats increase with temperature. The dissociation and ionisation processes, which absorb large quantities of heat, depend not only on the temperature, but on the pressure. The average molecular weight, w, also changes and modifies the gas law (page 30), which becomes

$$\frac{p}{\rho} = Z \frac{R^* T}{w}$$

where the factor Z, which remains unity until 200–3000 K, rises to 1.2 by 2500–6000 K (when oxygen dissociation is complete), increases further to 2 in the range 5500–12 000 K and, when all the atoms are singly ionised (above 12 000 K), it reaches 4.

At very high temperatures great care is needed in defining what is meant by temperature and how it should be measured.

The difficulty of evaluating flow quantities in a hypersonic flow arises because so many parameters are inter-dependent and also because, although the atomic changes occur very quickly, the time taken for the transfer of energy to settle down to new equilibrium conditions is sometimes of the same order as the time taken for the air to flow past the body.

Hypersonic experiments are also fraught with these transient effects. The heat liberated in a tunnel airflow at these speeds makes continuous running very difficult to engineer and many tunnels run only intermittently. Instruments have to respond to the airflow in less than a thousandth of a second, and a much larger number of measurements are needed to define a hypersonic flow than a subsonic one. The complementary technique of firing small models through still air is frequently employed in hypersonic experiments. Small models (1–12 in or 2–30 cm long) are fired from special guns or electromagnetic accelerators down a range past photographing stations. Successive photographs enable measurements to be made of drag, lift, stability, boundary layer condition and heat transfer (fig. 9). Models have been tested carrying small radio transmitters (fig. 14). Even the orbital re-entry of a body into the Earth's atmosphere has been represented by such a technique. The model 'bullet' is fired into a large bell mouth, down which air of varying density flows.

Free molecule (Newtonian) flow

In a highly rarefied gas, e.g. in an industrial high vacuum process or the upper atmosphere, the frequency of collisions between gas molecules becomes negligibly small, but there are still enough molecules to permit the definition of average pressure, density, temperature and mass velocity. At about 100 miles (160 km) above the earth, the mean free path length is about 10 ft (3 m) and there are 10^{13} molecules in each cubic inch (16.4 cm^3). Such a condition is typical of the 'free molecular flow' regime. This is sometimes referred to as Newtonian aerodynamics because near the surface of a body there is one stream of arriving particles crossed by another composed of particles bouncing off the surface. These streams do not interact, as shown in fig. 1. In a flow of this type, a boundary layer cannot exist.

Magnetohydrodynamic flow

In addition to the normal aerodynamic equations of flow describing density, pressure, viscous effects and heat transfer, an M.H.D. flow also requires the inclusion of forces and energy transfers encountered in conventional electromagnetism. As the fluid can conduct electric currents and these create magnetic fields, the following additional relationships apply which are most neatly expressed in vector form:

$$\text{Ampère's law:} \frac{\partial D}{\partial t} + J = \nabla \times H$$

D is displacement current, J is current density, H is magnetic field strength.

Faraday's induction law: $\dfrac{\partial B}{\partial t} = -\nabla \times E$

B is magnetic induction, E is the electric field strength.

Ohm's law: $J = \sigma(E + U \times B)$

σ is electrical conductivity and U is the air velocity.

In most problems $D = \varepsilon E$, $B = \mu H$ (ε = susceptibility, μ = permeability). $1/\varepsilon\mu$ = (velocity of light)^2, and $\nabla \cdot B = 0$. $\nabla \cdot D = \rho_e$, the charge density.

The Lorentz force is $F = J \times B$.

The Euler equation of fluid motion now has to include these effects and becomes:

$$\rho \frac{dU}{dt} + \rho(U \cdot \nabla)U = J \times B - \nabla p + \mu_1 \nabla^2 U$$

where p and μ_1 are the aerodynamic pressure and viscosity. M.H.D. effects can be shown in laboratory experiments with liquid mercury. Other examples are the English Channel, which conducts a total current of several thousand amperes, disturbing compasses, and the magnetism in the interior of the earth. M.H.D. in air has been called magneto-aerodynamics and is described in chapter 8.

Computing aerodynamic flows

The problem of defining the complete exact equations of aerodynamic motion and then solving them in particular cases has challenged mathematicians for two centuries. The most fundamental physical-mathematical formulation for fluid flow was first derived by C. L. M. H. Navier in 1823 and Sir George G. Stokes in 1845. The classical Navier–Stokes relations have an apparent but deceptive simplicity,[8] viz.

$$\frac{D\rho}{Dt} + \rho \nabla \cdot V = 0$$

$$\rho \frac{DV}{Dt} + \nabla p - \frac{1}{R}\nabla \cdot \tau = 0$$

$$\rho \frac{De}{Dt} + p\nabla \cdot V + \frac{1}{R}\left(\frac{1}{Pr}\nabla \cdot q - \tau:\nabla V\right) = 0$$

The vector V is the velocity with components $u_i (i = 1, 2, 3)$, e is the internal energy per unit mass; the tensor τ is the shearing stress, the vector q describes the heat flux. D/Dt represents the substantial time derivatives.

Even today with large digital computers which perform calculations about

100 million times faster than was possible in the last century these exact aerodynamic equations can only be applied in special cases. Ingenuity and special approximating methods are still essential and since the early 1960s there has been unprecedented progress in the application of numerical techniques to a vast number of fluid mechanics problems covering all the regimes listed in table 5. A new brand of aerodynamicist has appeared who can work closely with the computer whose power and speed is still increasing rapidly. Such an aerodynamicist working in one country can gain access to the world's largest computer in another continent by means of communication satellite links and landline connection. A typically large computer used in this way is ILLIAC IV built by Burroughs at Ames Laboratory of NASA. By using sixty-four parallel arithmetic units it allows simultaneous processing of sixty-four data sets. Such parallel processing requires the rewriting of existing sequential programs.

The types of flow which are now accurately soluble by computer include:

Inviscid flow over three dimensional aerofoils with bodies.
Transonic flow about simple wing shapes.
Three dimensional supersonic-inviscid flow fields for complex configurations with shocks present.
Two dimensional laminar boundary layers with and without compressibility.
Extension to three dimensional boundary layer flow is limited where the flow separates.
Many turbulent flows can be predicted reasonably accurately.
In small Reynolds number separated flows velocity and pressure distribution are reasonably accurate.
Some special cases of fully viscous unsteady flow.

Other examples are quoted elsewhere in the book, viz. meteorological forecasting (page 58), noise fields (page 26) high speed flows (page 154) and flow with combustion and chemical effects (page 42).

Flows which cannot yet be modelled satisfactorily by computer include transition from laminar to turbulent flow over a surface (and the subsequent relaminarisation in certain cases), unsteady shock wave-boundary layer interactions and even some practical inviscid supersonic flows between $M = 1$ and $M = 2$.

A few special examples of numerically calculated aerodynamic flows will suffice to illustrate this intriguing and recent development of the aerodynamic art.

Development of viscous layers over a multiple aerofoil

To increase wing lift when landing at slow speed, aircraft wings arrange extensions at front and rear to collect and deflect more air. Boundary layers are created on all surfaces which interact with each other. A program called MAVIS developed at RAE with industrial help in the UK solves this problem

Fig. 18 Development of viscous layers over multiple aerofoil

as shown in fig. 18. The agreement with experimental results is remarkably close; neglect of viscous effects gives a result which bears no relationship to reality at all.

Surface pressures over a complete aircraft wing in transonic flow

The flight speed represented was M = 0.825 and parts of the upper wing had regions of supersonic flow, aft of which there was a spanwise shock wave starting from near the root trailing edge out to the tip where it was at about $\frac{1}{4}$ chord aft of the leading edge. A finite difference relaxation solution of the small perturbation equation was used requiring 200 000 grid points to cover the wing. The calculated values of local pressure coefficient agreed better with full scale than did a wind tunnel test, especially the location of the shock which are attributable to Reynolds number effects at model scale. Inaccuracies behind the leading edge were due to neglect of some viscous effects.

Unsteady viscous flow

Most of the numerical methods employ a network or grid as described in chapter 1. Mathematical relationships identifying changes in flow quantities must be shown to be computationally stable and methods are assessed for precision. A desired accuracy is gained by the smallness of the grid and the shortness of the time interval. A somewhat different approach was developed by the Los Alamos Scientific Laboratory, USA, using Stretch, a large computer previously employed on problems of nuclear physics. Entitled MAC (Marker and Cell) method it represented the full Navier–Stokes equations and employs two sets of grids.[9] One is fixed within the domain of study but the other consists of cells which move with elements of the fluid. They then trace out the equivalent of a ciné picture of the moving flow. This novel technique proved highly successful in reproducing mathematically the very unsteady flow in the wake of a bluff object (fig. 19). The method has also been successfully applied to liquids which have a free surface. The marker cell is then able to distinguish which particles of fluid need the special treatment required at the boundary of the mass of the fluid.

Rarefied gas methods

The preceding examples refer to continuum flow and would not be applicable to a Newtonian type of flow. A computer technique called Monte Carlo is

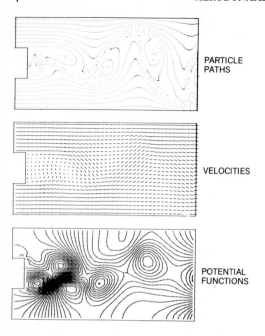

Fig. 19 The MAC Method. Viscid flow past a rectangle

used in such cases and individual particles are 'projected' through the volume of interest–appropriate forces deflecting its path and its point of arrival on a surface. The computation is then repeated many hundreds or thousands of times introducing random variations in the quantities as chosen by the computer from a set of random numbers. In the final result the effects are averaged from all the individual discrete calculations.

Such a brief review can do scant justice to this remarkable new aerodynamic tool which is available to scientists and designers.

There is no panacea however and the important feature is to add together iteratively both viscous and inviscid solutions about a body thus getting an acceptably good approximation.

Airflow measurement using laser methods

Light beams are used frequently in experimental aerodynamics.[10] At low speeds a slit of light introduced across the flow will illuminate vortices and breakaways if the tunnel is filled with smoke. At supersonic speeds when the air is significantly compressed or rarefied, i.e. at Mach Numbers exceeding 0.4 the change of refractive index is used to display shock waves, boundary layers and wakes by the shadowgraph (fig. 9) or Schlieren process. The advent of the laser, a coherent light beam of nearly a single wavelength, has led to its employment in many novel ways for flow exploration. A laser beam will be reflected by dust particles in the air whose velocity can be measured by means of the Doppler effect. This technique has been used in wind tunnels to explore the oscillatory airflow over a stalled wing at about 20° angle of attack, and also the vortex wake pattern of a helicopter rotor. A great advantage is the absence of disturbance to the airflow under study by the introduction of measuring tubes and supports. A disadvantage is that a foreign agent like dust sometimes has to be added to the flow to make particles more visible. The instrumentation and extraction of aerodynamic information from the raw data is complex and cannot yet be applied to all airflows of interest. Other experiments explored the interaction of a boundary layer with a shock wave–a notoriously difficult experiment with many different zones of flow in close proximity. Three applications in full scale indicate its wide range of usefulness: as a true airspeed indicator measuring several hundred metres ahead of an aircraft, so giving early warning of dangerous downdraughts or wind shears when approaching the ground; the measurement of large vortices in the wake of transport aircraft approaching a runway which are a source of disturbance to following aircraft; at longer range in determining pollutant particles high in the atmosphere, including volcanic ash. Holograms can be obtained across supersonic flow in wind tunnels from which shadowgraph, Schlieren or interferometric patterns can be produced (fig. 72). In plasma flow laser diagnostic probes were vital in assessing progress towards nuclear fusion power. It has even been proposed that the thermal velocity of a single atom could be measured by a laser time-of-flight velocimeter.

5
Natural Aerodynamics[11]

The air is delicate
                Macbeth: Shakespeare

In this chapter the atmospheric winds and disturbances caused by natural processes are described, as well as the occurrence of vortices, circulations, boundary layers and convective flows. Model experiments can help to explain these atmospheric flows and also to overcome problems in designing large buildings which must withstand the force of the wind.

Meteorology

Meteorology and weather forecasting clearly require an accurate knowledge of the motions and energy changes in the atmosphere. Geophysical fluid mechanics, which deals with the physical causes and effects of atmospheric movement, has received a great stimulus in recent years from two sources. The extensive measurements of hitherto unexplored regions now being made regularly by balloons, rockets and satellites give a much more accurate picture of the state of the atmosphere. Large capacity computers enable predictions of the future state of the atmosphere to be calculated from this data and from the six controlling differential equations governing pressure, temperature, density, heat energy, and viscous and inertia forces.

    All planetary atmospheres are influenced by three overall effects. Because of the force of *gravity*, the air density, pressure and temperature fall with height. The decrease of temperature is called the 'lapse rate' and the mean value is roughly 2°C per 1000 ft (6.5°C/km). The *rotation* of the planet takes the air round with it, so the air would tend to have a mean rotational velocity varying from a maximum at the equator to a minimum at the poles, were it not for other disturbing effects. The Sun supplies large quantities of *radiant energy* of many wavelengths during the daytime, so there are diurnal heating and cooling changes as the air passes from day to night and vice versa.

    Because the atmosphere has almost unlimited freedom of movement, there is a range of distinct flow patterns varying in size from the very largest which extend over thousands of miles, to the very smallest which take place in a scale of a few feet.

The different motions of the atmosphere are here described in order of size from the largest to the smallest.

Large-scale circulations

The simplified geographical idea of trade winds, westerlies and equatorial doldrums, tells only a part of the complex large-scale mixing processes. It is only in recent years that the way in which wind energy is transferred amongst these warm and cold wind zones has begun to be understood. One surprising phenomenon, discovered in 1946, is the jet stream (or planetary wave). It is an undulating 'river' of air encircling the globe, in which the wind currents exceed 100 knots (50 m/s). These are usually two streams, the subtropical and the circumpolar, and they change position and strength as they weave in between the major depressions and cyclonic air masses. Their position and strength are of great consequence to navigators of aircraft, and have to be allowed for in the flight of missiles and spacecraft. A representation of the general circulation of warm and cold air masses has been shown experimentally in cylindrical tanks containing two fluids of different density which do not mix. Their surfaces are subject to differential heating and cooling and they are rotated to represent the motion of the earth itself. Swirling flow patterns, traced out with aluminium particles, show frontal systems of the type seen in daily weather charts, and also longer-term effects similar to those unpredictable 'spells' of weather. Jet stream flows have been demonstrated in other model experiments with liquids.

Energy changes involved in large atmospheric motions are enormous. The total kinetic energy in the winds is estimated to be 10^{14} kWh (3.6×10^{20} J) and the energy converted in a single thunderstorm is 10^9 kWh (3.6×10^{15} J), the same as that of a megaton nuclear bomb. In a single snow storm (New York 4th February 1961) 40 million tons of snow fell, and to melt such a snowfall the heat of $2\frac{1}{2}$ nuclear bombs would be needed. These figures may help to allay fears of the effect of nuclear explosions on the weather! Surprisingly, the kinetic energy of the winds is only about 2% of the total solar energy reaching the Earth.

Large explosions and aeroclysms

There are other large waves which encompass the earth as a result of very large explosions. These are not gravity waves but involve the compressibility of the air as does a sound wave. The well-known curvature of sound waves from gun explosions is caused by the changes with height of the air density in the lower atmosphere. 'Big' bangs give detectable pressure rises at distances exceeding 600 miles (1000 km). An 'aeroclysm' rocks the whole atmosphere, sending pressure waves right round to the antipodal point and back to the place of origin. The Krakatoa eruption of 1883 was the only known aeroclysm until the large nuclear bomb tests of October 1961 which were proved to be aeroclysms from barograph records taken in Sweden. The Krakatoa explosion had an estimated energy release of 5000 Mt (5×10^{12} kg) of TNT.

Convective motions

If a volume of air, generally at rest, lies over a horizontal surface which is warmer than the air, there will be a flow of heat to the air and a free convection flow will be established. As warmed air becomes buoyant and rises from the surface, so other colder air from above will flow down to replace it. How does this up-and-down motion arrange itself? Is it entirely random, like turbulence, or is it periodic? The answer is that there is sometimes an orderly pattern of flow, and at other times an intermittent effect. In the regular pattern, air forms roughly polygonal cells, within each one of which there is a vertical circulating flow. The air rises in the centre, cools by radiation or other heat transfer mechanisms at the upper surface, flows laterally to the edges of the cell, and there descends to the lower, warmer layer. By this means the air velocities are the same on either side of the boundary between cells. In fact, the flow directions can be reversed, and the direction of flow depends on the variation of kinematic viscosity with temperature. Liquid sulphur, for example, exhibits opposite variations above and below 153°C (426 K).

This 'cellular' convection was discovered by Bénard in 1900 and is remarkable for its very wide application in size, temperature and type of fluid. It can be seen in tea-cups as the cold milk mixes with the hot tea, and the dirt polygons found in old snow lying in fields are believed to result from deposition of air pollution in a cellular convective process. The evolution of thermals (thermal currents or gradients) over a still ocean shows a cellular structure with dimensions of several hundred feet, as can be seen from seagulls' flight. Somewhat similar effects occur on the surface of the Sun, at temperatures near 7000 K in the highly radiating hydrogen and helium gases. Figure 75 shows the cellular pattern. The boundaries are here shown clearly as the flow at the edges is at a lower temperature and consequently less bright.

In the other type of convective flow, 'bubbles' of warmer air created over sun-heated land, sea or man-made objects, detach themselves from the surface and rise upwards. These help sail-planing, cause 'bumpiness' felt by aircraft, and eventually lead to 'streets' of clouds, thunderstorm 'anvil' clouds and give rise to energy transformations leading to lightning, and hail formation. Extreme examples of local convection are present in the development of the 'mushroom' cloud following a nuclear explosion.

Vortex motion

A circulatory vortex flow occurs in hurricanes (typhoons), tornadoes and dust devils. Hurricanes are very large circulating air masses which have extremely low barometric pressures at the core (eye). The wind speeds exceed 125 mph (56 m/s), material destruction on land is serious, and they may extend over several thousands of miles across the oceans. Tornadoes and dust devils, however, are much smaller, but are perhaps more spectacular; locally, tornadoes are at least as destructive as hurricanes.

Tornadoes. A tornado, or water spout, is a violent natural vortex reaching down from clouds to land or sea, nearly vertical at the ground but often horizontal higher up. Its general flow pattern is similar to the simple model shown on page 12 with a central 'core', outside which the velocity falls approximately inversely as the radius. The flow is complicated by strong vertical currents in the core reaching hundreds of miles per hour. There are conflicting theories about the details. Some maintain these are violent up-currents, other report downstreams. The starting of a tornado requires the simultaneous existence of several phenomena of which some important ones are vertical instability, high moisture content at low levels, wind shear effects above 10 000 ft (3 km) near a front and triggering action of hills or ridges in giving a starting disturbance. It is not suprising that the complex motions in the tornado are not completely understood, for the difficulties and dangers of experimental measurement are obvious, and much evidence was derived from secondary effects, such as photographs of the external shape at a distance, rapid falling and rising of the barometer as the vortex passes by, observations of damage to walls, telephone poles and girders, vertical height of water sucked out of ponds, erosion tracks across fields, and so on.

Velocities may be as high as 250 knots (130 m/s) in the core, which can sweep out a path of intense destruction from 5 to 50 yds (5 to 50 m) wide. In addition to these chaotic damaging effects, there have been several recorded instances of intense vertical up-currents in the core leading to fantastic 'lifting' effects. Men have been lifted hundreds of feet, a grain-binder a quarter of a mile, small items up to 40 miles (65 km) and, most remarkable of all, a pair of trousers with $95 in the pocket was blown 39 miles (63 km)!

Although not common in Europe, tornadoes have been observed there to extend from $\frac{1}{5}$ to 300 miles ($\frac{1}{3}$ to 500 km) indicating a surprising persistence. In the USA severe tornadoes have caused up to 600 deaths in a single path (1927), and an average annual damage of $14 million; sufficient incentive for considerable interest in a better understanding of these phenomena. The aim of research is two-fold: more accurate prediction of the occurrence and movement of tornadoes, and fuller knowledge of their aerodynamic characteristics to help perfect the design of tornado-proof buildings.

In the USA a large effort has been organised to measure the characteristics of tornadoes–how they form, grow and decay–in order to improve tornado warnings and understand better how their intense energy is focussed. A National Severe Storms Laboratory (NSSL) was set up at Norman, Oklahoma in the heart of the tornado belt, administered by the US National Oceanic and Atmospheric Administration (NOAA). This Laboratory uses special Doppler Weather Radar which can measure windspeeds throughout the whole region in which a tornado is developing from within a rotating thunderstorm. Unlike ordinary radar which can detect the position and strength of rainfall from the echoes returned by the raindrops, the Doppler Radar senses also the air velocity in cloud, or in clear air conditions. The Doppler Radar had to be specially designed for a most complex task. It had to range out to 285 miles (460 km) for general surveillance and reflectivity of rain and within a radius of 71 miles (115 km) determine wind speed up to a maximum value of 76 mph (34 m/s) with an accuracy better than $2\frac{1}{4}$ mph (1 m/s). These parameters were decided from observations of the general characteristics of typical tornadoes. Three measurements continuously recorded are (i) the echo power which gives water content and precipitation rate, (ii) the wind speed towards or away from the receiver, and (iii) the velocity dispersion, i.e. shear and turbulence. The radar system designed to achieve this has a power of 750 kW† operating at a wavelength of 10.3 to 11.1 cm at a frequency of 2.8 GHz. The derived outputs from the received signals are in digital format capable of driving computers, and are displayed on large TV screens and printed out on coloured charts. This is aerodynamic measurement on the grand scale and has already produced a new class of precision data of considerable help to tornado prediction. Severe storms have characteristic velocity patterns which the Doppler radar shows clearly. Meteorologists distinguish the scale of storms as follows: the very large weather systems are termed macroscale involving air masses exceeding 1240 miles (2000 km), the mesoscale refers to smaller scale events—meso-α 124–1240 miles (200–2000 km), meso-β 12–124 miles (20–200 km) and meso-γ 1.2–12 miles (2–20 km). Mesocyclogenesis is the term for a weather system which creates cyclonic storms which in turn may later give rise to tornadoes. Radar patterns are called signatures and mesocyclonic storm and tornadic vortex signatures have been observed. The latter may extend vertically up to 6 miles (10 km). This was particularly successful on 24 May 1973 during what must surely be the most comprehensively observed tornado of all time. General weather forecasts early in the day suggested the likelihood of a tornado and by 1300 hours all resources were mobilised. This included the radar, a mobile photographic unit with still and ciné cameras, balloon sondes from nearby areas, lightning strike measurements, meteorological aircraft and examination of surface damage. In fact a severe tornado lasting 26 minutes was in contact with the ground for a distance of 11 miles (17 km) at Union City, only 31 miles (50 km) from the NSSL itself. Details are in the 238-page report of the investigation,[12] and figs 20, 21 and 22 illustrate the tornado at its height, a diagrammatic representation of its growth and the Doppler radar tornadic vortex signature (TVS).

Accurate and comprehensive measurements of this type are an essential part of tornado prediction. Two other components are mathematical and physical modelling of the phenomenon. An example of the former is the complex numerical prediction methods for mesoscale convective pressure systems developed at the Environmental Research Laboratories of the NOAA at Boulder, Colorado, USA, by Fritsch, Chappell and colleagues. This sets out a model weather situation on a square 250 miles (400 km) a side with grid points every 20 km (smaller scale than that described on page 60) and starts the process with conditions typically conducive to the existence of a squall line. The equations of motion pay particular attention to convection

† Radio and radar quantities conventionally only in SI units.

NATURAL AERODYNAMICS 55

Fig 20 Tornado at Union City, Oklahoma

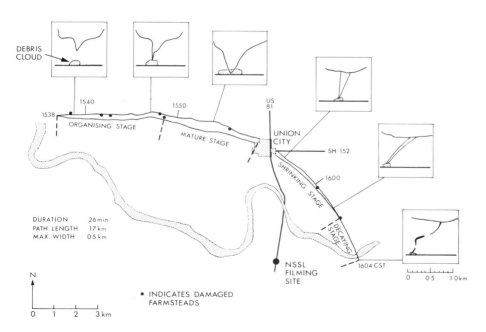

Fig. 21 Damage path of the Union City tornado with sketches of funnel and associated debris cloud as seen from the south.

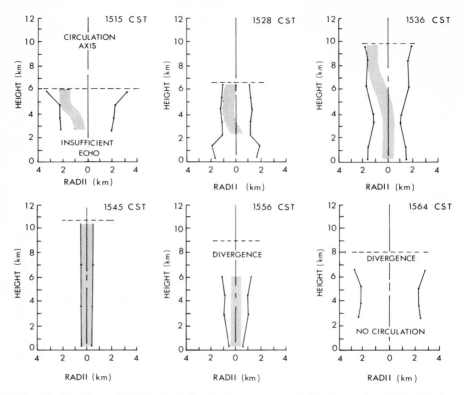

Fig. 22 Position of TVS (stippled) relative to centre of tilted core circulation. Dark dots indicate measured circulation radii. Dashed line identifies Doppler data collection upper limit in the circulation region. Tornado was on the ground from 15.38 to 16.04 hrs

flow and transfer of heat energy into moisture in the atmosphere. After relatively short representation of time quite violent storms develop; the programme can then be repeated with changes in some variables, e.g. artificial 'seeding' of storm clouds by injection of freezing 'particles'. These mathematical models confirm how complex are the atmospheric motions and how elusive are those small effects which seem to act to create that especially localised and violent rotating vortex, the tornado. Very significant, but remarkably simple, experimental models of a flow having many of the observed characteristics of a tornado were devised by Dr Stewart Turner at Cambridge. He rotated a vertical cylinder containing water and injected a stream of air bubbles into the centre so that as they rose they entrained an upward water flow behaving as does a rising convection in a cloud. A vertical vortex resulted which was very stable; it rose rapidly to the top and drew in more circulation from beneath so that it also grew downward. This is the exact characteristic of the tornado. A noticeable result was that the vortex would only form and persist for a small range of bubble flow rates. This sensitivity has also been confirmed by an equivalent mathematical model of

Turner's experiment conducted in Australia by Dr Roger Smith of Monash University. It looks as if this sensitivity to the creation of a steady vortex lies at the heart of the explanation of why so few thunderstorms actually give birth to tornadoes. Many of the existing means of tornado prediction are imprecise and often lead to false alarms; there is one comparison where a local forecast gave a 9 minute warning compared to 26 minute lead time made possible by the greater information provided by the Doppler. Even the latter seems small enough to take precautions or evacuate because of the uncertainty of the actual tornado path. Another complication is the double tornado one of which was well recorded and analysed at Elkhart, Indiana on 11 April 1965. The two cores rotated relative to each other, one growing in strength at the expense of the other. In the analysis reported in reference 13 a theoretical method for the tornado aerodynamics allowed for buoyancy in the core flow, surmised an influence of the upper jet stream affecting the upper boundary of the tornado and discussed an explanation due to possible electromagnetic effects in the core. A further investigation performed by NASA in 1971 measured residual magnetism in iron nails in damaged buildings along a tornado path as evidence of electrical flow. The belief that tornadoes form conductors of electricity between the earth and storm clouds and that electric heating of the air could be a source of the tornado energy has been studied for some time. Clearly the mathematical and experimental models so far described do not take account of M.H.D. effects (page 44) and should this indeed be a significant effect much further basic aerodynamic information will be required to understand the mystery of the tornado.

Hail damage

Hail damage to crops is a significant economic hazard in continental areas. Hail is formed by supercooled water droplets being convected upwards in cumulonimbus clouds at speeds of about 50–150 ft/s (15–46 m/s) to freeze in colder upper regions. Reduction of hail damage by seeding storm clouds with silver iodide from ground generators, aircraft and rockets is now practised in the USA and the USSR. American tests claim over 50% reduction in hail over the year 1957 in parts of Oregon and California. This technique is still in its infancy but it is likely to be increasingly used and to be joined by others such as distributing carbon black to change the heat absorption in upper moisture layers in the air and promote cloud formation. In all these instances knowledge of the motion of air, the carrying of small water and ice particles and the transfer of heat energy is essential to guide work along profitable lines.

The conventional wind-tunnel is not very suitable for demonstrating atmospheric convective effects, although some experiments have been made where the tunnel floor was heated and the ceiling cooled. Rotating water tanks have already been mentioned and a great deal of work has been done with a simple 'cloud-box' apparatus to study cloud and radiation fog formation.

Weather forecasting (also called climate prediction)

There are generally three types of forecasting associated with increasingly long time scales. Firstly there is the daily radio or TV forecast for tomorrow and probably the next two or three days. Secondly there is a medium term prediction from a month up to a year ahead and lastly a much longer global guess at the trends of change in the atmosphere as a whole over periods of decades in such matters as temperature, pollution content, carbon dioxide level and thickness of the ozone layer. The first is essentially a problem in fluid dynamics; the other two also involve thermodynamic processes and statistical inferences. The introduction of global atmospheric computer modelling has indeed transformed the first, improved the second notably and is also used in the third, not only in predicting long periods into the future, but also back into the past for comparison with existing records in order to test alternative hypotheses for sensitivity to random variables. In this section the modus operandi of the atmospheric computer models will be described and how data is acquired from weather satellites in orbit. Energy and pollution effects are dealt with on pages 186 and 189 respectively.

The weather equations

The Global Atmospheric Research Programme (GARP) associated with World Weather Watch is a wide ranging international programme of the World Meteorological Organisation which was formally inaugurated in 1967. This required (a) improved physico-mathematical models of the global atmosphere, (b) a global observing system to provide adequate weather data, and (c) adequate computing facilities to process this data and means to test the adequacy of existing methods with a view to their improvement.

Sir John Mason, Director-General of the UK Meteorological Office published in 1971 the equations of atmospheric motion[14] used for GARP as developed by Bushby and Timpson (1966) and Benwell and Timpson (1968). They adapt the equations of velocity, continuity, energy and state by using pressure height as a measure of vertical distance. With this simplification the five governing equations for dry air become:

$$\frac{\partial u}{\partial t} + u\frac{\partial u}{\partial x} + v\frac{\partial u}{\partial y} + w\frac{\partial u}{\partial p} = 2\Omega v \sin \phi - g\frac{\partial h}{\partial x} + F_x \qquad (1)$$

$$\frac{\partial v}{\partial t} + u\frac{\partial v}{\partial x} + v\frac{\partial v}{\partial y} + w\frac{\partial v}{\partial p} = -2\Omega u \sin \phi - g\frac{\partial h}{\partial y} + F_y \qquad (2)$$

$$\frac{\partial w}{\partial p} + \frac{\partial u}{\partial x} + \frac{\partial v}{\partial y} = 0 \qquad (3)$$

$$\frac{D}{Dt}\left(\frac{\partial h}{\partial p}\right) + \frac{w}{\gamma p}\frac{\partial h}{\partial p} = \frac{\gamma - 1}{\gamma}\frac{1}{pg}\dot{Q} \qquad (4)$$

$$h = \frac{R}{g}\int_p^{p_0} T\frac{dp}{p} \qquad (5)$$

x and y are easterly and northerly directions.
u and v are horizontal components of the wind components along x and y.
$w = Dp/Dt$ is a measure of vertical speed.
h is contour height of a constant pressure surface.
Ω is the Earth's angular rotation, and ϕ is the latitude.
F represents viscous, turbulent and frictional dissipative forces per unit mass, viz. the transfer of momentum and heat from one layer to another by eddies on a smaller scale than is represented by the grid of points in the computer.
g is the acceleration due to gravity.
γ is the ratio of specific heats of air at constant pressure and constant volume.
R is the universal gas constant.
p is the atmospheric pressure.
$\dot{Q}$ is the rate at which heat is added to unit mass of air.

To include the effects of evaporation, condensation and precipitation of water three other equations are required. The atmosphere is divided into a number of layers separated by surfaces of equal pressure of thickness h' and the transfer of water between layers lead to a restatement of equation (4).

$$\frac{\partial h'}{\partial t} + u\frac{\partial h'}{\partial x} + v\frac{\partial h'}{\partial y} + w\frac{\partial h'}{\partial p} + \frac{wh'}{\gamma p} = \frac{\gamma - 1}{\gamma g}\frac{\Delta p}{p}(\dot{Q} + ML) \qquad (6)$$

Here Δp is the pressure difference across the thickness h', M is the mass of water vapour condensed per unit mass of air, and L is the latent heat of condensation.

The two other equations take account of the continuity of moisture between levels, considering both rainfall and water vapour.

If a layer is unsaturated, its humidity-mixing ratio r (mass of water vapour per unit mass of air) changes by horizontal and vertical advection, and by evaporation of water falling into it from above. Thus:

$$\frac{(\partial r)}{(\partial t)_{\text{dry}}} = -\left(u\frac{\partial r}{\partial x} + v\frac{\partial r}{\partial y}\right) - w\frac{\partial r}{\partial p} + M_a \qquad (7)$$

where M_a is the rate of transfer of liquid water from the layer above. If, once the layer reaches saturation, its excess moisture is deemed to condense and fall out as rain into the layer below, we may write:

$$\frac{(\partial r)}{(\partial t)_{\text{sat}}} = \frac{\partial h'}{\partial t}\frac{\partial r}{\partial h_{\text{sat}}}, \quad M_b = \frac{(\partial r)}{(\partial t)_{\text{dry}}} - \frac{(\partial r)}{(\partial t)_{\text{sat}}} \qquad (8)$$

where M_b is the rate of transfer of liquid water into the layer below. The rate of condensation of moisture in the layer is then $M = M_b - M_a$, the latent heat released, ML, being allowed for in the thermodynamic equation (6) and its dynamical consequences computed.

These equations are then fed into a large computer together with data on wind speed, pressure, temperature and humidity measured by radio sonde balloons, etc. Physical quantities are derived for 12 000 grid points, 31 miles (50 km) apart, covering an area around the UK and most of Europe of size 4000 miles (6500 km) by 3100 miles (5000 km). Ten atmospheric levels are computed from 1000 mbar (100 kPa) to 100 mbar (10 kPa) [approximately at sea level and 53 000 ft (16 000 m)]. The mathematical model also allows for the effects of mountain ranges such as the Alps and those of Wales, Scotland and Scandinavia, for the frictional drag of the land and sea on the air, for horizontal eddy diffusion and for the influence of convective clouds. To maintain stability of the computations the equations are integrated every 90 s so that a 24-hour forecast involving 10^{10} numerical calculations takes 30 minutes on an IBM 360/195 machine.

These computer models have made some remarkably accurate comparisons, e.g. over a period of two weeks of the development of cyclonic weather systems commencing in Texas, USA, and moving across the Atlantic to finish east of the UK. Daily detailed forecasts of rainfall and prediction of depressions and frontal systems up to three days ahead over the whole Northern Hemisphere are now commonplace.

How adequate are such mathematical models and how could they be improved? A great shortcoming of the model is the lack of provision for partial cloud cover, because clouds are much smaller than the grid lengths so that vertical motion on the cloud scale cannot be represented. The convective motions which lead to cloud formation and subsequent radiation of energy from the cloud tops provides a very powerful negative feedback system and hence a stabilising action. It is probably because of the neglect of the cloud effects that some models suggest very wide divergences of the future climate and gloomy predictions of new ice ages or thermal catastrophe. How can one judge such possibilities?

Over the years models are checked against real weather situations and in this way a background of verification is built up. The mathematical model itself suffers from 'mathematical diffusion' which is a feature of all grid models which smooth out all short curvatures of the functions stored in it.

Weather in the public mind is characterised by extremes which habitually break records established over years, decades and even a century. How often is it announced that such a circumstance has never happened since regular meteorological records were started? A few instances will show what kind of accuracy of predictions might ultimately be expected of the mathematical models:

1953	Most extensive floods of this century and probably longer. Floods over England and Holland due to combination of winds in the North Sea, movement of a cyclone and high spring tides.
1961	Heavy rains in East Africa, lakes rising to highest levels in the twentieth century.
1963–4	Driest winter in England and Wales since 1743.
1964	Heaviest and widest snowfalls in South Africa in June since 1895.

1964–5   Supposedly ice-free port of Murmansk blocked by ice.
1968–73 Severest drought ever experienced in African Sahel and Ethiopia.
1969   Lowest frequency of westerly wind days in UK since 1785.
1976   Great heat wave in Europe. Longest drought in England. Temperatures in England exceeded a 300 year record by 4°C.

Some of these are short term events such as exceptionally heavy snowfalls; others such as droughts, ice blocking and flood conditions, imply longer term changes persisting perhaps for several years. Although some of the inadequacies of the present day models are known and may well be allowed for in the future with bigger computers and more precise data inputs (particularly of the ocean conditions, for example) it is not obvious that, even then, such severe extremes will be anticipated. Are there any longer term changes to the atmosphere caused by man-made heating, pollution, nuclear fall out, etc. which might be at work in a very subtle way? Before attempting to answer these questions it would be helpful to take stock of the contributions to observational meteorology from satellite systems.

Satellite meteorology

Clearly the large weather computers require prodigious amounts of data and although this is fairly readily obtainable over inhabited land masses by means of balloons or rocket sondes, it is not easy and is very expensive to provide the same coverage over vast tracts of the Earth, especially above the Southern oceans. The weather satellites, which were first launched in 1960, conveniently fill this important need. The earliest types, such as TIROS (TV and Infra-Red Observation Satellite), traversed over the atmosphere in orbits inclined to the equator and at a height of 300–500 miles (482–804 km). They were stabilised so that they continuously pointed the TV cameras and infra-red sensors towards the Earth beneath and recorded information on tape recorders for later transmission as they passed over a few ground receiving stations. Whole families of other satellites have been employed since then using better sensors, orbits and means of data transmission. NIMBUS followed the ten TIROS satellites at three times their weight; TOS was superseded by ITOS (Improved TIROS Operational Satellite) and by 1970 TV cameras were replaced or supplemented with high resolution radiometers and microwave spectrometers. A knowledge of the English language is inadequate to understand the world of weather satellites: a dictionary of acronyms is also required to decipher the names of satellites, the special sensor techniques used (VHRR, SR, VTPR, SPM, TOVS, ARGOS, etc.) and the organisations operating them (ESSA, NOAA, ESA, ISRO, etc.).

Two special orbits are important for meteorological work. From 1972–1976 four ITOS spacecraft were placed into circular almost-polar orbits at 932 miles (1500 km) altitude. Regular pictures were obtained of clouds, sea conditions and ice fields, including both polar zones. The other is the geostationary orbit at 22 200 miles (35 800 km) above the equator where the satellite's speed around its orbit is such that it completes one revolution every 24 hours. Once stabilised in such an orbit the satellite then 'hovers' over a fixed place on the Earth's surface giving a fixed observatory in space. For the Global Weather Experiments from 1 December 1978 to 30 November 1979 four such weather satellites were employed. GOES 1 (Geostationary Operational Environmental Satellite) launched in October 1975 was stationed over the longitude of the eastern US; GOES 2 in June 1977 over Central America and GOES 3, June 1978, over the longitude of the western US. These American craft were joined by the European Space Agency's (ESA) METEORSAT 1 and the Japanese HIMAWARI 1 over the Pacific. To fill a gap created by the delayed Russian spacecraft, GOES 1 was later moved around its orbit to cover the Indian Ocean.

The achievements in the 1960s were to make available large quantities of clear photographs of the World's atmosphere showing as never before the development of cloud systems, hurricanes, tornadoes, squall lines, thunderstorms, mountain waves and icebergs. A remarkable service called APT or Automatic Picture Transmission enabled ground radio stations anywhere in the World to receive weather pictures from the satellite as it passed overhead. One such satellite, ESSA 8 (Environmental Science Services Administration), operated successfully for seven and a quarter years transmitting no fewer than 265 136 photographs to APT terminals. All the even numbered ESSA satellites offered APT service; all those with odd numbers transmitted global weather information to the US Department of Commerce for processing and distribution to the major US meteorological centres. The most recent advance in this technique is carried by the NOAA–TIROS N (National Oceanic and Atmospheric Administration, USA) which is to be the workhorse metsat of the 1980s. Its AVHRR or Advanced Very High Resolution Radiometer provides images on four channels for cloud, snow, ice detection and for surface temperatures with a resolution of 2.5 miles (4 km). For the High Resolution Picture Transmission service this is improved to 0.6 miles (1 km) by using 13 ft (4 m) parabolic receiving antenna.

But photographs (fig. 23) although useful, are not enough to provide the more precise data needed for the weather equations given on page 58. What precision is needed and how is this obtained? It has been stated (reference 14) that the following accuracies were required for the global atmospheric computer models envisaged for the GAR Programme.

Table 6 Global atmospheric model accuracies

Air temperature to	1 K
Atmospheric pressures	0.3% (3 mb, 300 Pa) at surface, 0.3 mb (30 Pa) at 100 mb (10 kPa)
Wind velocities	± 7 mph (3 m/s)
Vapour pressures	± 10%
Sea surface temperatures	within $\frac{1}{4}$°C
Cloud cover	± 20%
Precipitation intensity	± 10%

Fig. 23 Cloud streets indicate the flow of Polar air. The wake near the centre is caused by the disturbance of Jan Mayen Island

Atmospheric temperature and humidity are deduced from measurements obtained by several types of radiometer. The radiation emitted by a layer of uniformly mixed gas in local thermodynamic equilibrium is directly dependent on its temperature. The difficulty is that radiations of many wavelengths are created at all levels of the atmosphere and some arising from lower levels are partially absorbed by upper air layers. At first sight it would appear to be an insuperable task to decipher such measurements being recorded several hundreds or thousands of kilometres above the earth. It can only be done by using several sensors of particular wavelength or frequency which respond differently and quite sharply to different levels of the atmosphere. How this is done can be demonstrated by a classic comparative experimental analysis performed by NASA on NIMBUS-5 on 22 January 1973. Satellite results were compared with radio sonde and ground radar meteorological data to assess errors. The NIMBUS-5 carried five different radiation sensors, viz.

ESMR   Electronically Scanning Microwave Radiometer for earth and atmospheric radiation at 19.35 GHz (1.55 cm). This penetrates cloud.

ITPR   Infra-red Temperature Profile Radiometer contained:
–four channels in the 15 μm CO_2 band,
–one channel in the 19 μm rotational water vapour band,
–one channel in the 11 μm atmospheric window (i.e. it sees through the atmosphere as if it were transparent),
–one channel in the 3.8 μm atmospheric window This sensor derived the temperature profile from the surface up to 84 500 ft (25 km) with a vertical resolution of 11 ft (3.5 m) and an estimated accuracy of 2 K.

NEMS   Nimbus-E Microwave Spectrometer
–three channels tuned to oxygen lines near 60 GHz,
–one channel tuned to the 22 GHz water vapour line.
–one channel tuned to the 31 GHz atmospheric window.
NEMS can sense atmospheric radiances even in the presence of clouds.

SCR   Selective Chopper Radiometer.
This has four optical filters each containing four channels covering wavelengths from 2.06 μm to 133 μm. This permits sounding up to 148 000 ft (45 km), again to a precision of 2 K.

THIR   Temperature-Humidity Infra-red Radiometer is a scanning bolometer.
–one channel 10.5–12.7 μm window night and day; surface and cloud tops,
–one channel for 6.5–7.2 μm water vapour absorption which senses integrated moisture content of the upper atmosphere.

It is one thing to receive such a rich array of data: it remains to sift out the temperature and humidity profiles as a function of atmospheric pressure. This is done by means of atmospheric weighting functions which are very different for all the sensor channels described. Each channel has a sharp peak at one pressure level hence most of its radiation will be received from this level; other channels cover other heights. The following table summarises this effect.

Table 7 Atmospheric weighting functions (NIMBUS-5)

Atmospheric pressure (mb)	Channel	Maximum transmittance rate	NEMS-2 transmittance at this mb
850	ITPR-4	0.61	0.06
600	NEMS-3	0.52	0.30
450	ITPR-3	0.52	0.46
250	NEMS-2	0.53	0.53
90	ITPR-2	0.40	0.25
87	NEMS-1	0.68	0.22
30	ITPR-1	0.31	0.04
8	SCR-2	0.56	0
1.5	SCR-1	0.40	0

1 mb = 100 Pa.

Note that the peak transmittance rates cover the whole pressure range and how rapidly the value falls on either side of the peak value. When these weighting functions are correctly applied to all sensor signals a temperature profile results. These, when compared with radio sonde data, indicated an overall average error of −0.9°C with extreme values of +6.1°C and −6.7°C.

Atmospheric winds may be measured by time lapse pictures of certain cloud systems taken by geostationary meteorological satellites. But this is somewhat haphazard and gives no indication in clear air. Pressures are now regularly measured by balloons equipped with precise radar altimeters accurate to ±33 ft (10 m). Those flying at a pressure level of 100 mb (10 kPa) fly for many months and make many circuits of the Earth. But the lower levels are of great interest too but here balloons suffer from ice, frost and snow and do not survive for long.

Considering the extremely complex nature of the atmospheric motions it is perhaps not surprising that collecting and analysing meteorological information is an extremely involved matter. Moreover the quantity of data produced is so vast that it cannot be taken in mentally on a daily basis. The weather satellites are very sophisticated but will not totally replace ground based radar, radio sondes, balloons and other observational techniques on land and sea. It is not yet clear whether Laser or Doppler Velocimetry as described earlier will be usable for extensive measurements of global weather–it will probably be confined to use in special local regions for supplementary investigations of aerodynamic processes.

The biosphere[15]

In treating some of the longer term influences on the atmosphere, its constituents and its dynamic behaviour, it is convenient to introduce this concept, first named by the Austrian Eduard Suess in 1875, and subsequently developed by the Russian mineralogist Vladimir Vernadsky in the 1920s. Defined as that part of the Earth where life exists with the special features (i) where water exists in large quantities, (ii) input of energy (primarily from the sun), and (iii) there are important interfaces between liquid, solid and gaseous states of matter. In our present meteorological context the last is of major relevance. The interchanges between atmospheric gases, the sea and the land are subdivided into major cycles, viz. water, oxygen, carbon, nitrogen and sulphur. The energy cycle describes the driving power of the solar radiation and how it warms the surface, creates winds which drive ocean currents, evaporate water from seas, rivers and lakes, and in very dry regions creates dust storms. The large scale circulations ensure that the atmosphere is well stirred up and the changing concentrations of gases and particles are thoroughly mixed.

The earth is estimated to contain 0.36 billion cubic miles (1.5 billion km^3) of water subdivided into the oceans (96.4%), ice (2.8%), fresh water (0.7%) and surprisingly water vapour a mere 0.0005%. The world-wide evaporation and precipitation rate is nearly 39.4 in (100 cm) per year. Over the oceans the total annual average precipitation lies between 42 in (107 cm) and 45 in (114 cm) with evaporation greater (because of river inflow) of 46 in (116 cm) to 49 in (124 cm). Overland average precipitation is 28 in (71 cm) with corresponding evaporation of 18 in (47 cm). The total rainfall water delivered to the US annually is about 1440 cubic miles (6000 km^3) from water vapour passing over it which is ten times greater. The oceans carry warm and cold currents and these modify the air circulation over them, changing humidity and cloud cover. The most renowned is the steady Gulf Stream in the North Atlantic, but there are less welcome large gatherings of cold oceanic water which occasionally collect off coastal regions and promote unusually wet spells for some months at a time. The oxygen cycle is concerned with the exchanges between living organisms on land and sea promoted by photosynthesis. Exchange cycles of long period have been estimated for these processes. An amount equal to all the Earth's water is split by plants and reconstituted again after a period of two million years. Oxygen associated with this process cycles every 2000 years and atmospheric CO_2, after being liberated by plants and animals, is once again fixed by plants after 300 years. The carbon cycle is similarly involved with the biomass but additionally is deposited in sedimentary deposits of coal, oil and gas. Since a major concern at present about the CO_2 cycle is the effect of burning fossil fuels in the atmosphere this subject will be dealt with later under man-made effects. The nitrogen cycle quantifies how much nitrogen is fixed naturally by atmospheric processes and is liberated by igneous rocks; it is recycled by vegetation but more slowly than oxygen; it is a generally passive constitutent of air and of less consequence to meteorology.

Overall measurements of quantity and cycle change times derived from study of the biosphere lead naturally to the last topic in weather forecasting to be discussed, i.e. that of predicting longer term climatic changes. It is as if the daily weather forecast is looking at the 'grain' in a photographic emulsion which cannot be seen when observing a moving picture or TV screen for it is the long term change of overall image pattern that is significant and observable. In order to assess the sensitivity of the overall atmospheric circulation to major changes of primary parameters computer models have been employed. The present models inadequately represent the effects of clouds in radiating away solar energy and the effect of ocean surface temperature. Nevertheless they have shown remarkable realism in demonstrating the effects of seasonal shifts of temperature and monsoons.

Many studies have investigated the overall effects produced by changes in solar radiation, soil moisture, vegetable cover, ozone and man-made effects such as dust and pollution. One such study by Wetherald and Manabe of Princeton changed by 2% the solar energy input. A rise of 2% produced an increase of 3°C in mean global surface temperature but a correspondingly large decrease lowered temperatures as much as 4.3°C. Snow cover and radiation were markedly affected as was precipitation which was changed 27% by a 6% change of solar input. Other studies relating to the drought region of the Sahel in Africa showed that provided a region was initially wet it would create local climates tending to make the moisture persist. In an investigation of the particularly cold winter of 1962–3 in the UK it was observed that the eastern tropical Atlantic Ocean was 2.5 K warmer than normal. The model showed that such a change produced a low surface pressure west of the Bay of Biscay and increased pressure off Greenland which modified the wind flow across the UK and explained the anomaly.

In this section the new meteorology has been described with its large capacity computer modelling and remote sensing from above the Earth's surface. Meteorological satellites are now an established routine and during the 1980s there will be about twenty sophisticated measuring satellites continuously in orbit. Much has still to be learnt at the fundamental level however. General improvement of precision with smaller networks of computed points are required and more precise physical relationships are needed at the molecular scale, e.g. radiative transfer of CO_2 and ozone, water vapour and clouds. At a micro scale are the turbulent exchanges in the planetary boundary layer of heat momentum and water vapour, and on a larger scale still the laws governing frontal circulations, mountain disturbances and the deep tropical convective system and hurricane formation.

Terrain aerodynamics

The streamlined flow of a generally steady wind over cliffs, hills and rocks can be calculated, some results being shown in fig. 24. The gravity waves formed in the lee of mountain ridges extend upward to heights many times greater than the mountain. Gliders have been assisted up to over 30 000 ft (9 km) in altitude by such atmospheric waves. The exact form of the waves depends on the thermal conditions and on a resonance effect between wavelength and hill shape. 'Rotors' (see fig. 24) create unexpected down-draughts. Calculation of these streamlined flows has been helped by flow experiments in salt water of varying density.

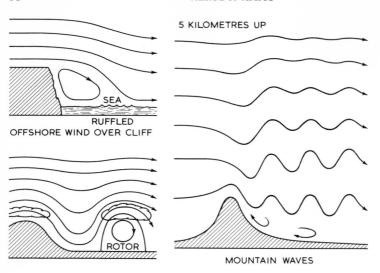

Fig. 24 Terrain aerodynamics

Other unpredictable downdraughts occurring over the Rock of Gibraltar caused serious embarrassment to pilots and constituted a real danger. Fig. 25 is of the 1/5000th scale model of the Rock of Gibraltar tested at the National Physical Laboratory to measure vertical air currents in the lee of the rock near the airfield landing strip. Flow directions were indicated by wooltufts and, in spite of the very small scale, the agreement with actual conditions was satisfactory. Similar wind-tunnel tests made earlier on a model of Mount Fuji in Japan were not so reliable, as the smooth contours of the mountain allowed laminar flow on the model, where the full-scale flow would have been turbulent. As the Rock of Gibraltar surface is rough and uneven, the correct turbulent 'boundary layer' could be expected on the model.

Turbulence

Turbulence is that unsteadiness and gustiness of the wind which we observe everywhere, blowing smoke from bonfires, 'flying' flags on poles and buffeting the sails of yachts. This stirring motion, though very evident on the surface of the earth, goes on throughout the whole atmosphere. It is a difficult motion to define because of its very random nature, which means that the path of any one particle of air cannot be predicted for any appreciable time, and yet it possesses statistical quantities which can be measured. These show that turbulence is the basic mechanism whereby several qualities of the atmosphere are diffused, viz. the heat, water vapour, momentum, and suspended smokes or pollens. The general turbulence acts as a damper to the disturbing effects of convection, frontal systems and jet streams, thus helping to reduce differences of velocity between neighbouring regions.

Fig. 25 Flow exploration over Gibraltar model

It is to be expected that weather prediction will prove more successful as the science of geophysical fluid mechanics evolves, but it is unlikely that satisfactory long-term forecasts will be possible. The major difficulty is the triggering effect of small disturbances which lead to large-scale motions. An example of this is the nocturnal low altitude jet stream, only discovered in 1953, which is found between 800 and 2000 ft (240 and 610 m). It builds up in the night and is believed to explain partly the birth of storms. Its presence is now thought to have caused some hitherto unexplained flying accidents.

Micrometeorology

This term describes the behaviour of the atmosphere from the ground up to the first few hundred feet, where it is considerably affected by the surface of the earth. In some ways the air in this region behaves like a boundary layer and there are empirical laws expressing the change of mean wind-speed with height, as described, for instance, by Davenport of the NAE of Canada (fig. 26).

Coming closer to the ground, there are many phenomena of considerable practical importance to human life now receiving a great deal of scientific attention. Chief among these are the movement of water by transpiration from plants and evaporation from lake surfaces, the blowing of soil and sand and the driving of snow. The aerodynamics of these processes involves viscous effects, turbulence in the air and even the lift-drag ratio of dust particles. Many wind-tunnel and other model tests have been carried out with success, and many specialised centres of research have been set up, one being at Nottingham University which has a climatological wind-tunnel used particularly to study the water economy of plants and grasses. Other examples are wind- and water-tunnel experiments of smog formations in mountainous regions near the coast, convection currents assisting the flights of locust plagues and the design of aerodynamic filters to prevent the ingestion of wind-blown sand into the intakes of aircraft engines.

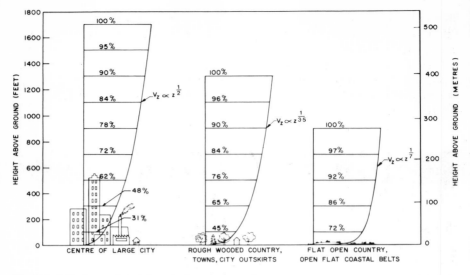

Fig. 26 Wind speed: height relations over surface

The science of micrometeorology evolved mostly in secrecy in response to the military need of predicting the dispersion of poison gases in war. It is in a way ironic that this science now helps mankind to mitigate the harmful effects of pollution from industrial fumes and smoke which can reach poisonous concentrations in adverse weather conditions.

This is a fascinating corner of aerodynamics and a lot of useful research still remains to be done. The experimental apparatus is quite inexpensive and the economic returns appear attractive. A few examples are given in the following paragraphs.

Soil erosion

The loss to agriculture from soil erosion can assume very serious proportions. 300 million tons (300 Tg) of soil were removed in a single dust storm in the USA in 1934. Prevention of soil erosion and conversion of desert regions to areas of cultivation require a better understanding of the mechanics of soil erosion. Many wind-tunnel tests have been made of different-sized top soil grains to determine in particular the most effective wind breaks. Even the oscillating lift and drag forces on representative soil grains have been measured with strain gauge balances. Large-scale drifting in deserts, and wind breaks to improve water conservation in oases, are also subjects of study.

Snow[13]

Superficially, wind-driven snow may appear to have basic mechanisms similar to those of soil. The flow patterns are not exactly similar, however, for the size and shape of the particles, as well as their texture and temperatures, are different. Differences in the type of terrain, vegetation, geometry of roads and living quarters between hot sandy deserts and snowy regions also influence the kind of problems that arise. In wind-tunnel tests, snow has been represented by flaked mica, sawdust, and gypsum, and scaling factors to be satisfied by a model 'snowflake', taking into account its elasticity, have been worked out. At 1/10th scale, Borax is a satisfactory model.

The effect of snow drifting on transport vehicles is illustrated in fig. 27.

Fig. 27 Snow deposit on wind-tunnel bus model

A surprising example of snow control has been a proposal to set up wind breaks to promote a large level snowdrift adequate for an aircraft runway.

Evaporation

A law of evaporation from water surfaces depends on the Reynolds number:

$$N = aR^b$$

where $N = Ex/\Delta C v_e$,
E = evaporation weight per unit area and time,
$\Delta C = C_i - C_0$,
C_i = saturation water vapour concentration at T_m,
C_0 = water vapour concentration of ambient air,
T_m = mean temperature of surface over length x,
x = downstream distance from beginning of surface,
v_e = molecular diffusivity coefficient for water vapour into air.

(a and b vary with differently shaped surfaces and states of boundary layer, etc., but for a circular lake a good approximation is $N = 0.256\ R^{0.87}$.)

Wind driven circulations

Water currents in lakes influence the chemical and biological changes which control the distribution of sediments and polluting agents. The primary mechanism for these currents is the wind but it was not known how the wind energy was transmitted from the water surface into the main lake water down to the bottom. In a thorough series of analytical studies R. T. Gedney and colleagues at NASA Lewis Research Center at Cleveland, Ohio, solved the hydrodynamic equations and applied them to circular and rectangular lakes with and without a central island. Firstly steady winds were assumed and later eddies were included representing gusty variations of wind velocity over the lake. The variations in wind velocity were shown to have a marked effect on the water currents developed in the depths of the lake. This type of study was relevant to some of the acute pollution of the Great Lakes of North America.

Spraying of crops

Spraying with various insecticide and fungicide liquids is now a major agricultural process. Aerodynamics enters into the problem in the design of atomising nozzles to create droplets of the correct size (chapter 6), and the distribution of droplets around plant leaves and trunks by natural winds and turbulence. Spraying from aircraft and helicopters assists the work extensively, but there are other problems and particularly the harmful dispersal of insecticides downwind.

Forest aerodynamics

Wood grown commercially in forests is a vital resource for energy and construction materials, and the global need grows 4% per annum. Improved understanding of the effect of air movement in forests contributes to better utilisation of this precious resource. There are two aerodynamic aspects.

The first is concerned with the efficiency of growth, i.e. the rate at which the wind speed evaporates water from the leaves and this in turn has an effect on the uptake of soil nutrients and the exchange of atmospheric gases. Wind velocity also affects leaf temperature and the rate of carbon fixation which is basic to the production of woody material. Many measurements have been made in experimental forests of water uptake and evaporation in relation to rainfall, sunlight, windspeed and temperature. Unexpectedly, it was discovered that significant moisture was absorbed by the leaves of certain species in conditions of high humidity or fog. Even without rainfall in the strict sense, the trees can continue to grow. Hydroponics, or growing plants in the laboratory without soil with special nutrients, provided yields over ten times better than in nature. To improve productivity in the forest itself, research into forest growth is seeking more modest improvements, perhaps only 2–4 times, which would nevertheless make a significant economic impact on forest products. The partitioning of growth between different parts of the tree was also found to depend on the general wind direction.

The other aerodynamic effect is wind damage to forests, which can be economically disastrous. In 1953, for example, one severe gale in Scotland blew down four million trees, nearly five times the number felled annually in the whole of Great Britain. The attack on this problem was made on three fronts. Firstly, live specimens of trees of spruce, scots pine, douglas fir and western hemlock were placed in a large wind tunnel at the Royal Aircraft Establishment (RAE), United Kingdom, and the wind drag was measured. As wind speed increased, the frontal area of the tree shrank so that the drag force increased approximately as the speed, and not as the square of the speed (fig. 28).

The drag coefficients varied for different kinds of tree, viz:

Table 8

	C_D at 30 knots
Spruce	0.57
Scots pine	0.415
Douglas fir	0.37
Western hemlock	0.25

These tests also measured the uprooting forces exerted on the base of the tree trunk, by different gale speeds. Secondly, model forests were represented in another wind tunnel at the National Physical Laboratory by conical wire mesh models, having the equivalent scale drag of the real tree, and buffeting loads were measured at the base of the model trees by strain gauges.

The model tree, being of fixed shape, could not respond to changes of windspeed like the real tree. The standard model was designed to produce an average drag coefficient of 0.4 at 30 knots (15.4 m/s) with an area reduction of 0.56. The tests were then done at a constant speed. The models created the right kind of turbulence in the wake behind the tree, and other tests showed that Reynolds number changes did not noticeably affect the results.

The 'forests' were made in different patterns using standard model trees scaled from small bushes to very large mature trees. The regular plantation had 256 standard trees 'planted' at fixed distances apart. Curved margins, thinned margins, dense margins and wedge profile margins were also represented. One forest had irregular sized trees, and in another a ride clear of trees was represented.

These tests confirmed the observation that a gale striking the sharp edge of a forest rises sharply upwards, becomes turbulent and strikes downwards

Fig. 28 Testing the aerodynamics of a tree in the RAE 24 ft (7.32 m) tunnel

some way into the forest, creating violent oscillating forces on the trees. The new designs of forest margins tested, in which smaller trees were used in the first few rows made the air flow up and over without severe turbulence. The third aspect found that improved drainage allowed the roots to hold more firmly in the soil, and better withstand the buffetting of the wind.

Aerodynamic effects of the wind on man-made structures

Whereas vehicles designed to move through the air are streamlined to avoid energy losses in the airflow, man-made structures on the ground are anything but streamlined and the flat surfaces and sharp corners of buildings and bridges create severe forces, pressures and oscillations. There are two distinct issues; one is to be able to predict the local pressures and the total forces of the wind and then to make buildings strong enough to withstand them. The other is to deal with the exaggerated and gusty winds created around buildings and over ships. There is a special kind of aerodynamics which deals with the flow past bluff and sharp bodies. First, a closer look at the planetary boundary layer (fig. 26) as it affects man-made structures. Friction at the surface caused by trees, buildings, bridges, etc., creates turbulent eddies of many sizes which induce energy down from the main flow above. This layer is typically 2000 ft (600 m) high and is also marked by a horizontal change of the wind direction caused by the slowing down of the boundary layer friction and the Coriolis force. This effect is known as the Ekman spiral (after V. W. Ekman, 1905) and can be as large as 10–30°. The frequency of the wind velocity variations near the surface has been measured by Van der Hoven who showed that there were two predominant frequencies containing most of the wind energy, viz. the macro-meteorological element centred on a 4-day period and the micro-meteorological occurring at a frequency of about one per minute. Between these frequencies there is a virtual energy gap. This is important since the mean wind effects can be represented in experiments separately from the turbulent properties.

An adequate representation of the boundary layer has proved to be of crucial importance in model experiments; before this was done model results, when scaled up to full size, bore little resemblance to those actually measured. In the wind tunnel the lower surface layers of air are slowed down by a series of obstructions such as toothed barriers, grids, roughened floors and tapered spires. The variations in eddy size are introduced by elliptic shaped wedges which induce vortices of different sizes. A good illustration of the effect of boundary layer over the ground surface is the streamlined flow pattern ahead of a vertical circular cylinder. Without a boundary layer there is a stagnation line with the main flow coming to rest there and then parting to flow either side, all the flow being essentially horizontal. With the boundary layer present a large swirling vertical eddy forms ahead of the cylinder so that the stagnation line is lifted clear of the surface. One result of this is a marked

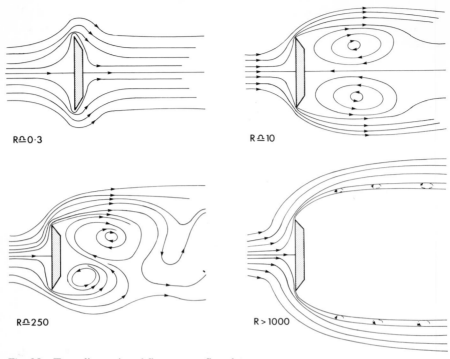

Fig. 29 Two-dimensional flow past a flat plate

downward flow to the surface along the front generator of the cylinder (somewhat as in fig. 31).

Bluff body aerodynamics are characterised by severe breakaways at edges and corners. The subsequent wake is turbulent and very dependent on Reynolds number. Figure 29 shows these effects; simplified to a two-dimensional example.[16]

Wind loads on structures and buildings

There have been several spectacular failures of bridges by excessive wind loading, notably the Tay bridge in 1879 and the more subtle destruction by aerodynamic oscillation of the Tacoma Narrows suspension bridge in the USA in 1940. Suspension bridges had been blown down before, but the exact mechanism of the oscillation was not understood. The Tacoma bridge was quite large and had vibrated on several occasions before the actual disaster. The complex oscillations involving both heaving and twisting motions grew to amplitudes of several feet, and the world's attention was drawn to the event by a film showing the final collapse of the whole structure. Winds are the dominant force ruling many engineering constructions, but accurate measurements have only been introduced in recent years. At the turn of the century, unreliable empirical formulae were evolved based on the collapse of shop windows and changes in barometer pressure readings during the passage of storms.

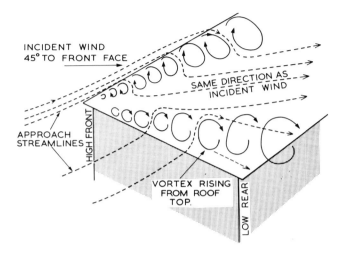

Fig. 30 Double conical vortex flow over flat roof

Very severe peak suction pressures occur on flat-roofed buildings, and an explanation for this can be found in the creation of vortices at the edges as shown in fig. 30, which are similar to those of Concorde. Measurements of pressure over small areas can easily be made in wind-tunnels, by means of small holes flush with the surface. The flow over sharp edges is not sensitive to Reynolds number and thus small models are usually reliable. While this may not present any particular problem for conventionally-shaped houses, which are catered for by good building practice, buildings of unconventional shape should always be subjected to wind tunnel explorations especially when untried cladding material, fastenings or substructures are employed. Apart from local pressure, the total force may give rise to excessive stress on foundations.

Skyscraper designs have been subjected to some notable aerodynamic research, for here the wind loads and their variation with time can cause tremendous forces. The total airload on the Empire State Building in New York in a 100 mph (45 m/s) hurricane would be about 2000 tons (20 MN). Model tests made at the time of its construction (1930) were incorrect because the boundary layer effect was not properly represented. In Canada, the very unconventional Toronto City Hall building consists of two crescent-shaped towers facing each other, with a narrow gap between the two edges. Wind tunnel tests on a model revealed torsion and bending stresses far exceeding those calculated by standard methods.

The wind loads on buildings are affected by local ground contour and the disturbed wake of other structures. Measurements are made for design purposes of the steady air loads on many kinds of girder beams and assemblies. These loads are exceeded in the wake of another bridge or building and allowance should be made for this. Failure to do this was one of the causes of a spectacular collapse of three large concrete cooling towers at Ferrybridge U.K. in 1965. Eight were mounted together close enough to create blockage and interference effects. Prediction of the wind loading on these hyperboloid–conical shapes was found to be sensitive to the exact profile of the shell. Aerodynamic lift was enough to overcome its deadweight and put the windward concrete structure in tension. The buffeting wake also broke down another tower at the rear of the others. The internal airflow rising upwards in the tower at much less than gale speeds mixes with the external flow in an unsteady fashion creating additional stresses. Lessons learned from this experience contributed to a new squat design of tower–1500 ft (460 m) in diameter, but only 300 ft (91 m) high–adequate to cool a generating station having a power output of 2.7 million h.p. (2000 MW).

The aerodynamics of the Rock of Gibraltar were also subject to further model tests in 1975 to establish why rain collecting surfaces were destroyed by winds lighter than those for which they were designed. Large vortices created by the delta-shaped rock were shown to be responsible–a further explanation also of the gustiness experienced by aircraft pilots landing at the airfield.

Tornadoes cause such severe damage to buildings, that tornado-resistant design requirements are not mandatory in current building codes or standards in tornado-prone areas but on the other hand the risk of a strike is very low. However, a great deal is now known about tornado wind effects and how to cope with them so that there are adequate design rules for constructions that dare not risk tornado damage, e.g. nuclear power stations. A Disaster Potential Scale subdivides tornado/cyclone winds into five bands from 74 mph (33 m/s) to above 155 mph (70 m/s) listing likely damage to signs, windows, roofs, small and large buildings. In estimating the effect of wind pressures acting on a building as a tornado passes by there is a model of a tornado wind flow. This relates to maximum translational and rotational velocities, pressure drop, and its rate of change and tornado size. Because of the rapid pressure changes the flow of air in and out of a building will relieve the external pressure loads by setting up balancing pressures inside. The airflows are computed with a mathematical model, taking account of the size and location of vents and any special passages and 'connectivity compartments' built in for this purpose. Methods also permit estimates to be made of the velocity attained by solid objects picked up by the tornado and which cause severe impact damage. A wooden plank, for example, could reach 130 mph (58 m/s) in a Type III tornado and a car weighing 4000 lb (1810 kg) could achieve an incredible speed of 131 mph (59 m/s) in the severest Type I.

Interest also centres on the flow of air through buildings and up chimneys, for this is directly relevant to heat conservation, ventilation and warming schemes. Vertical temperature distribution affects the magnitude of these effects. The normal atmospheric temperature decreases roughly by 2°C per 1000 ft (305 m)–this is termed the lapse rate. If a region of air has a higher negative lapse rate than this it will rise by buoyancy, and vice versa. Chimney efficiency is markedly influenced by variations in lapse rate which are difficult to represent in model tests. The effects of large explosion waves, and the rapid changes of pressure as they sweep past buildings, form yet another class of phenomena which has been the subject of important research.

Wind-induced discomfort around buildings

There has been an increasing tendency over many years to build very large box-like buildings with generous spaces and 'precincts' around them to induce a feeling of spaciousness and light. During the 1960s it became noticeable that several of these large planning schemes were plagued by exceptional high local winds, severe gustiness, vorticity and the inconvenience of dust and rubbish. A new aerodynamic phenomena was at work, well illustrated in fig. 31, of a simplified arrangement of one large slab-sided building with a smaller

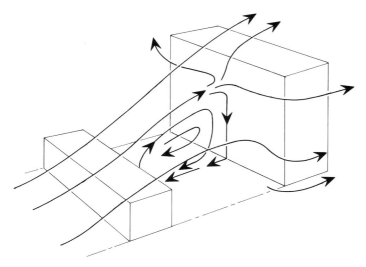

Fig. 31 Idealised flow pattern around precinct buildings

one upwind of it. Notice that the stagnation point is close to the top of the large building resulting in a very high speed downward flow to ground level and a large vortex bridging the gap between ground level and the flow leaping over the smaller building and rising up to clear the large one. Wind speeds at the ground can be three or four times greater than those hitherto experienced in older type towns. Model tests have explored the ground level speeds as functions of the relative dimensions of the buildings and there are several model tests of actual precinct areas all of very individual properties and shapes. Since also the wind flow patterns change dramatically for different directions of wind it is impossible to predict the effects at the architect's scheming stage. Model testing is now an essential part of the planning process but the large box-shaped buildings seem inevitable wind raisers whatever proportions are used. Although much research has identified scales of acceptability and discomfort for ground wind speeds (gustiness is taken into account too), many of the cures are unsatisfactory, e.g. building large roofs over shopping precincts. In some cases shrubs and trees can be strategically placed to tame the wind but such treatment could hardly be expected to harness the vertical gale descending directly down a large skyscraper.

Similar undesirably strong surface winds are to be found on high speed ships or hovercraft. Murphy reported in 1974 some interesting model experiments on a retractable porous barrier that could be raised at the windward end of a ship's deck. A porosity generally of 50% rising to 100% in the top quarter of the barrier led to reductions in wind speed of 40–90% over a large deck area.

Wind-induced oscillations

Wind flowing past a structure can cause oscillation of many kinds. The wind speed, shape and structural stiffness, and damping are all contributory factors, but there are four main types. In a steady wind, aerodynamic oscillation can be set up by the alternate *shedding of vortices* from a bluff body. A more complex oscillation involving an unstable non-linear lift characteristic of asymmetric shape is termed '*galloping*'. If the structure deflects under wind loads so that aerodynamic and structural forces interact in an unstable manner, it is said to *flutter. Buffeting* occurs when an unsteady wind subjects a structure to forces varying in speed and direction at a frequency disturbing to the structure.

In the last twenty years, however, these problems have been effectively tackled by wind-tunnel tests and the markedly reliable interpretation of their results. After the Tacoma Narrows bridge collapse, which was caused by vortex shedding from the edges of the flat side plates of the bridge, special wide wind-tunnels were built to test complete models of suspension bridges. These tunnels, which were over 50 ft (15 m) wide, were too specialised in function and are no longer in vogue. Instead, adequate evidence is obtained by testing models of sections of bridges (fig. 32) in standard wind-tunnels. The models should represent most of the constructional detail such as girders, platforms and handrails, and have the correct distribution of mass and stiffness. Reynolds number is not especially important, and testing can be simplified by having separate models for flexing and torsion. The models are tested over a range of wind speeds and any tendency of the model to oscillate on the spring suspension is observed.

Chimney stacks, TV masts and space rockets

The wind flowing past a vertical cylinder sheds vortices alternately from the sides. As each eddy forms, it causes a circulation 'lift' force sideways on the body, and as it is shed the force falls and the next is established, but in the

Fig. 32 Wind-tunnel test of model bridge

opposite direction. The frequency, f, of shedding of such pairs of eddies is given by the Strouhal number:

$$S = \frac{fD}{V}.$$

where D = diameter of cylinder, and V = fluid speed. S depends on R, but for circular cylinders $S = 0.2$ for

$$10^3 < R < 2 \times 10^5.$$

If this frequency is close to the natural frequency of bending of the cylinder and its structural damping is low, severe oscillations may result and lead to excessive structural loads. The most common remedies are 'strakes' or tall narrow flat plates attached to the sides to cut up the eddy flow, additional mass at the top of the cylinder to change the natural period of oscillation, and restraining cables from the top of the tower to the ground. These are not made taut, but have enough play to allow a small oscillation to build up which would be checked long before it grows to damaging proportions. Television transmitter masts often have aerials of peculiar shape, whose behaviour cannot be calculated but which can be tested in wind-tunnels. Long slender space rockets suffer from the winds at exposed launching sites. The Vanguard satellite launching vehicle encountered dangerous oscillations which were reduced by fitting strakes. Some difficulties might be expected in the much larger rockets such as the Saturn boosters, 200–400 ft (60–120 m) high, which have far smaller safety factors than any bridge and which, because of their light structural weight, are relatively much more flexible.

'Galloping' of cables

The periodic shedding of vortices from cylindrical cables is the origin of the well-known 'singing' of telephone wires, and of the Aeolian harp. The sound frequency and hence the pitch is that given by the Strouhal number for a given diameter and wind speed. Electric power cables stretched between towers several hundred feet apart have been observed to oscillate violently over amplitudes exceeding 10 ft (3 m) in a wide range of wind speeds. This cannot be explained by vortex shedding which requires a resonance closely dependent on a particular wind velocity. 'Galloping' instability is the term used to describe this effect, and it is caused by the deposition of sleet and ice on the circular cable building up an asymmetric shape. The criterion for instability has been derived by den Hartog to be:

$$\frac{dL}{d\alpha} + D < 0$$

where L and D are lift and drag forces respectively, and α is the incidence. Tests have now established which forms of cable are prone to instability; the semi-circle with flat side facing the wind is the most unstable shape known. If a cable of unstable shape begins to move across the wind direction its motion changes the incidence of the wind, and if the den Hartog condition applies, a lift force is produced in a direction to continue the lateral motion of the cable. This persists until the cable elastic tensions exceed the lift force and the cycle starts in the opposite direction.

Jodrell Bank, large aerials and circus tents

The 'Space Age' has brought with it not only the now familiar large cylindrical rockets, but large radio aerials of very unusual shape. Some are fixed, such as the large scatter antennae that bounce radio messages from ionised layers in the upper atmosphere, and the early warning missile detectors which are designed to withstand gales of up to 120 knots (60 m/s) in exposed areas, even though they may also be covered with many inches of frozen snow. The total calculated windloads are several hundred tons. Other aerials are rotated as necessary and have to remain steady in the presence of wind forces which not only vary from day to day, but also change with the angle of direction to the wind. Oscillatory flow over panels and edges is also important in its effect. Because such structures are mostly of very unconventional shape the wind-tunnel has proved the only means of measuring the flow forces. The resulting estimates are then used in designing the strength of the structure and the power of the electric driving motors and control gear. The Jodrell Bank Radio Telescope in the UK was the subject of such aerodynamic experiments. Other tests have been made on circus tents, a large restaurant at the top of a tower in Holland, and large flimsy ornaments such as those that are strung across streets at Christmas.

Fires, fog dispersal and smoke

Fire involves the combusion of a substance with oxygen of the air, and the heat thus liberated leads to an aerodynamic motion, usually with evolution of smoke. A great many problems of research are involved, as a brief summary of present-day activities will indicate. Factors influencing the spread of *accidental fires* throughout buildings and districts are being investigated by a co-operative wind-tunnel programme shared by a number of nations, and the result should be better fireproofing design and improved methods of combating fires.

Wartime bombing raids started large conflagrations with such an intense upflow of combustion gases that the induced sideflow reached hurricane force and made extinction impossible. Attempts were made during the war to *disperse fog* at airfields by burning oil and thus re-evaporating the water droplets. Novel heating and airflow model tests were done but the method had limited success and has never proved economically feasible. Factory chimney *smoke* often leaves the chimney in a random way, licking down the sides and depositing dirt. Redesigning the lip shapes, however, can change the airflow and reduce this effect. The efflux from smoky chimneys can be very unpleasant for the surrounding districts, and aerodynamic tests have been made with models to determine the correct height of chimney and velocity of efflux to minimise this nuisance.

On a larger scale, scientists are interested in the long-term effects on our atmosphere from the combustion of coal, oil and petrol and the generation of carbon dioxide. It has been estimated that 360 000 million tons (365 Pg†) of CO_2 have been added to the atmosphere by man's burning of fossil fuels, increasing the concentration by 13%. This progressive rise in the CO_2 content of the air has influenced the heat balance between the sun, air and oceans, thus leading to small but definite changes in surface temperature. At Uppsala in Sweden, for example, the mean temperature has risen 2°C in 60 years.

Air conditioning in nature

Apart from the transpiration of atmospheric gases and moisture through the surfaces of leaves in plants and trees, there are some intriguing examples of animals harnessing aerodynamic forces for survival.

The black tailed prairie dog found on the Great Plains of North America has extensive burrows underground, often 10 ft (3 m) deep and 50 ft (15 m) long, which require oxygen for respiration and a flow of air for ventilation.[17] It has been calculated that any natural convection through the tunnels would be inadequate for one animal in hot summer months, let alone a colony. It was observed that the prairie dog meticulously built up one of the two entrances to its burrow system to form a raised lip. Originally believed to be a

† Peta gram = 10^{15}g.

solution to prevent flash flood water entering the burrow, it is now established that the funnel-like entrance is the basis of a natural air ventilation system. The other entrance hole is a more gently rounded dome. As the wind sweeps across the prairie, the raised funnel lip forces the air to rise up, so increasing its speed. By the Bernoulli relationship, the air pressure is reduced and a suction is created. The airflow over the other entrance does not create such a depression and hence, air is drawn into it, through the burrow and exits at the raised funnel. The prairie dogs are careful to rebuild the entrances to the special design after rain or snow. Smoke measurements made by D. L. Kilgore of the University of Montana showed the existence of the flow of air which, although only 1.6 ft/s (0.5 m/s), was sufficient to change the air inside the burrow every 10 minutes. Wind tunnel experiments on models of several kinds of burrow opening made at the correct Reynolds number confirmed the differential pressures created by the different shapes. A difference in height was helpful as well as shape, and the prairie dogs knew enough to combine shape and height in the best combination.

It is interesting to speculate how the dogs hit upon this very energy efficient air conditioning arrangement. Presumably, the evolutionary effects worked on accidental differences and those families with better ventilated tunnels survived better than those without them. Without any formal aerodynamic training the prairie dogs must have had a keen sense of observation, and a fierce determination to survive. One can imagine other methods they might have tried, such as rushing through the burrows with fur extended to act as a piston. Such a method would be very costly in food energy terms, and hence would be quite unacceptable in the very strict energy conservation discipline that nature imposes.

A rather more subtle use of air flow is employed by the turret spider. His vertical burrow, created in dry sandy regions, is topped by a 'turret' made by a raised edge of sand covered by pieces of vegetation held together by silk. Surface wind passing over the turret which is typically 1 cm high creates a suction which induces an airflow from the surrounding sand soil into the burrow, and this brings with it water vapour from the saturated layers below the surface. This moist airflow prevents desiccation of the spider, and keeps the warm surface air from descending through the turret.

Termites build very large nests in which colonies of 2 million insects survive in very hostile environments. A whole colony of one African species can die after half a day's exposure to dry air and individual specimens cannot survive more than 5 hours. In 1953 Rahm and Lüscher made a classical study of the microclimate of the nests of five different termite species. The most advanced aerodynamic control system was found in the species Macrotermes, whose mound is typically 16 ft (5 m) high. The air conditioning requirements are very clear: a temperature around 30°C, a humidity between 96 and 99%, a supply of oxygen and an elimination of carbon dioxide. The two million inhabitants need 264 gallons (1.2 m^3) of air each day. Moisture is obtained from burrowing underground to a depth of 130 ft (40 m), and from fungi cultured in the nest. A difficulty is that with such highly saturated moisture the particles of the structure become impervious to gas flow, but the termite nest has developed a sophisticated way of overcoming this. Many air passages are provided which allow a continuous circulation of air vertically upwards through the centre of the colony where it gains heat, moisture and carbon dioxide. A cavity near the top permits a radial outflow to a returning series of vertical passages running close to the outermost part of the wall. Here the descending air dries, cools and exchanges carbon dioxide for fresh air. With each circulation, oxygen concentration increases 10%, so 10 circulations a day are needed to maintain equilibrium. The airspeed is about 0.1 in/s (22 mm/s) in the outer channels, reducing to 0.006 in/s (0.15 mm/s) in the nesting cavities. The researcher pointed out that this is less than the airspeed encountered by a termite when walking about, and hence the overall design avoids unpleasant draughts! It is not clear exactly how the nest adjusts to periods of continuous heavy rain when the normal aspiration of gas and cooling would be inhibited. It is believed that the occupants will open and close special channels in the surface at such periods–in other words, the whole structure self adapts to survive.

The aerodynamics of beehives is also full of fascinating natural control systems. Here again there are limits to temperature, moisture and carbon dioxide; the heat balance depending on the temperature and the metabolic rate in the hive. For cooling purposes 'ventilating' bees line up in the entrance and by tilting their bodies tail-up, create a combined draught that increases air circulation through the hive. By this means, bees are able to survive wide variations in external conditions which explains their wide geographic distribution. The termite system on the other hand is less adaptable, and is only to be found in certain tropical regions.

Passive cooling systems in buildings

Modern houses and particularly multiple storey business blocks employ powerful energy consuming heaters, refrigerators and fans for air conditioning. In comparison with animal systems just described, such prodigal waste of energy is to be deplored. Renewed attention is now being paid to some ancient methods employed in Middle Eastern arid regions to permit natural air conditioning of houses without recourse to external energy, apart from that of sun and wind.[18] Perhaps the most successful and enterprising of these methods are to be found in Iran, as part of its classical architecture with an antiquity stretching back 5000 years. There are five basic elements: the domed roof, the air vent (mounted atop the dome), the wind tower, the cistern and the ice maker. The domed roof with vent is the easiest to understand, for in principle it is akin to the prairie dog burrow entrance. The dome forces the air over it, increasing speed and decreasing pressure, which then draws air up and out from the room beneath. Fountains and pools within the room cool the moving air.

The wind tower operates in a more complex fashion, depending on wind speed and whether it is day or night. They can rise 115 ft (35 m) above ground and contain many vertical passages connected to several vertical slits created around the sides at the top. Wind flow enters the windward side slits, descends to the rooms beneath and returns to the tower, leaving by the downwind slits. At other times, e.g. of low wind speed it acts as a chimney conveying warm air from below. The cistern is a more complex combination of dome, vent and wind towers to promote a vigorous cooling air circulation over a large tank of water let into the ground. This becomes very cold during winter months, and is kept cool during the summer by the continuous evaporative cooling airflow. It is drained by a tap at its base, to which access is provided by a stairway. The ice-maker employs the near freezing conditions of winter nights to freeze the upper surface of long shallow ponds sheltered from day-time solar radiation by protective walls. Ice so collected can be stored in a cistern arrangement for use in hotter periods.

The effectiveness of the wind tower can be increased by separating it from the building and conveying the airflow by an underground tunnel which can either contain an underground stream of water or which has porous walls which are moistened from irrigation water supplied to the surface of the ground (fig. 33). Unsaturated (dry) air descending the tower will be cooled as

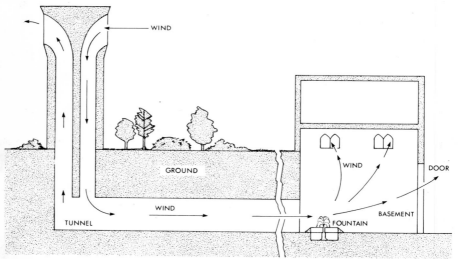

Fig. 33 Natural cooling of buildings in Iran. (From 'Passive cooling systems in Iranian architecture' by M. N. Bahadori. Copyright © 1978 by Scientific American, Inc. All rights reserved.)

it gives heat to the wall, and will be further cooled as it encounters the wet surfaces. The dry air evaporates the water, and in doing so is cooled further.

It is intriguing to realise that the distinctive architecture of Iran, together with the pools and fountains are a tribute to a very successful energy conservative means of air conditioning buildings in a very arid environment and which was developed many, many centuries ago.

6
Transport and Industrial Aerodynamics

A sense of burning fills the air—
  The Electrician is no longer there
                    Belloc

The previous chapter dealt with the effect of the wind on fixed man-made structures. In this chapter two applications of aerodynamics are described, which, although very dissimilar, have one common feature: they occur near the surface of the earth. This defines the air density, and as airspeeds are generally much lower than the speed of sound, ideal flow aerodynamics proves to be quite suitable. There are some exceptions in industrial aerodynamics where the airflow in pipes is often supersonic.

TRANSPORT AERODYNAMICS

Motor cars, trains, ships and bicycles create aerodynamic disturbances as they move, and the magnitude of the air forces and pressures is such that they have an appreciable effect on the vehicle's behaviour.[19] The major complications are the force of air drag retarding motion, the stability of bodies at high speeds, ventilation of passenger vehicles and snow and rain deposition on windscreens. Yachts are also included in this category because in spite of the fact that they depend on the natural wind for their power they are man-made and move.

Motor vehicles

Designing the exterior shape of cars to reduce air drag was introduced as 'streamlining' to racing cars in the late 1920s, and private cars in the 1930s. Its application was not at first always based on engineering necessity, but at present-day speeds, when air drag takes at least as much horse-power as the rolling friction (each about 25%), streamlining is essential. Major motor companies have large wind tunnels capable of testing full-scale cars at speeds of up to nearly 100 mph (45 m/s) and a large amount of data has been accumulated from wind tunnel tests.

A new impetus to reduction of air drag has been given by the increasing price of fuel and it is now customary to see drag coefficients quoted in descriptive brochures (often to three significant figures!) alongside values of top speed, fuel economy and novel engineering advantages. Whether this is function or fashion is questionable but there is undoubtedly a much greater public awareness of the aerodynamics of the motor car. The kind of steady, if slow, progress over the years is indicated by the average C_D of 0.7 for US cars of the late 1920s, reducing to 0.5 by the end of the 1940s. In Europe an average C_D for eighty-six popular makes was as high as 0.46 even in the 1970s. The range of values was from 0.37 to 0.52.

But drag is by no means the only aerodynamic feature of significance for road vehicles, although it is undoubtedly the most important. The overall flow field, consisting of streamlines, separations, wakes, vortices, interference with the road surface and rotating wheels, surface pressures, noise and rain effects must be understood comprehensively for many reasons. Stability at high speed, particularly the response to side winds and gusts can be deficient especially in low built streamlined sports cars. An efficient supply of cooling air is needed to remove surplus heat from the radiator. Intakes for internal ventilation, heating or air conditioning have to be placed where the dynamic air pressure is favourable and cannot ingest exhaust fumes. The chemical nature of the exhaust gases results from the combustion processes with air in the engine (see pollution in chapter 9), and within the engine itself are many interesting aerodynamic phenomena, e.g. the carburettor, air inflow into and out of the cylinder, the flame front and pressure rise during explosion and the aerodynamic drag of rotating parts.

The early streamlined designs of car attempted to create a shape that was like a half-teardrop with very rounded front and gently tapering rear. The latter had to be impractically long to effect a worthwhile drag reduction and was often directionally unstable. Almost all moderately sized cars now have a distinctly short afterbody characterised by definite wake flows and vortices. Two types of afterbody flow are shown in fig. 34. The recirculating bubble or spiral vortices determine drag, stability and the deposition of dirt, thrown up by the rear wheels in wet conditions on to the body and/or rear window. The drag coefficient is quite sensitive to the angle of the back window as in fig. 35. The lower drag of the fastback is marked and accounts for this increasingly popular, although rather angular, shape.

Care is needed in quoting drag coefficients for motor cars. Tests done by General Motors on a $\frac{1}{4}$-scale model fastback measured a drag coefficient of 0.27 at a Reynolds number of 700 000, which decreased to 0.23 at $R = 2$ million. Often tunnel models of new cars are simple shapes without practical details represented; a typical list of the drag increments from such items is shown in the table.

The skill of the designer is shown by the way he can blend together these many aerodynamic requirements. A good instance of this is the Lotus Elite GT four seater. It is low and wide and the top of the windscreen is more than half way back from nose to tail. The curved wedge-like forebody was kept low by having retractable headlights and a canted engine block; the resulting aerodynamic drag power is only 40 hp (30 kW) at 100 mph (45 m/s). A

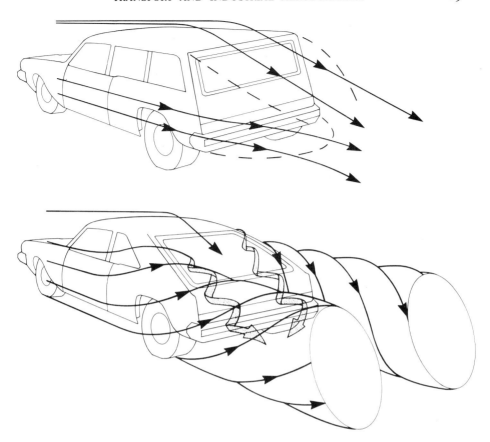

Fig. 34 Two types of afterbody flow

Table 9 Approximate contributions to total drag of a typical saloon car

		C_D	% (of 0.45)
1	Basic streamlined body	0.080	18
2	Ground effect on above	0.020	4
3	Adaptation of body to conventional styling	0.150	33
4	Wheel and wheel arches	0.090	20
5	Underbody details	0.045	10
6	Headlamps, bumpers, etc.	0.040	9
7	Engine cooling system	0.025	6

serious consequence of this found in experimental models was a large upward lift force on the curved forebody which decreased adhesion of the front wheels so becoming dangerous in crosswinds. The cure was to fit a wide narrow scoop under the front of the engine, which collected high speed air

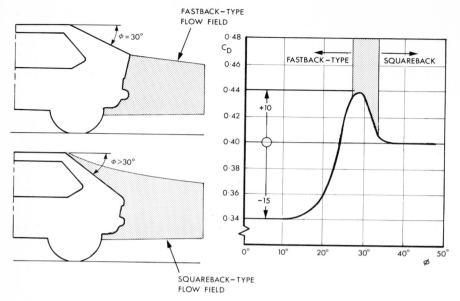

Fig. 35 'Squareback' and 'fastback' configuration effect on drag coefficient

and passed it through a shallow, wide radiator so that the retarded air was ejected into the boundary layer on top of the curved forebody thereby breaking the lift suction. In this way drag, stability, lift and cooling airflows were all combined successfully.

Commercial vehicles present a quite different class of aerodynamic flow. Because of the need for a large volume for carrying a variety of goods their basic shape is essentially that of a flat sided box. Overall dimensions are restricted and so there is little opportunity for streamlining as developed for the small private car. The Volkswagen bus of the 1960s presents a good example of the economic advantage to be gained from model-tunnel testing. The original design had sharp front edges, and wool tufts along the sides showed the flow completely broken away and turbulent. Even quite modest rounding of all the front edges and corners streamlined the airflow and reduced the drag by 40%. The fuel saving on all Volkswagen buses in service at this time corresponded to 130 000 tons (132 Gg) per annum. The new shape was presumably more expensive to manufacture, but aerodynamic noise was probably reduced.

Once again the oil price rise of the 1970s has recently focussed increased attention on ways of reducing drag on commercial vehicles. In today's terms 'Energy' is emphasised more than 'Power', e.g. in the UK by reducing freight vehicle drag by 10% would curtail fuel requirements by as much as 300 000 tons (304 Gg) per year. A comprehensive aerodynamic research programme at the College of Aeronautics, Cranfield, UK, has recently reported some unexpected solutions and results.[20] Many kinds of commercial vehicle shape were tested at scales of $\frac{1}{4}$ to $\frac{1}{10}$ in an 8 × 6 ft (2.4 m × 1.8 m) wind tunnel at maximum speeds of 100 ft/s (30 m/s) and up to 20° angle of yaw. In this way not only the air drag of new shapes could be measured but also any instability in crosswinds. By fairing over the cab of a small van reduced its C_D from 0.75 to 0.55. Even the fore and aft position of a box load on a flat bed truck was shown to affect the drag; when mounted immediately behind the cab the drag was 16% less than when it was loaded on to the rear end. Retractable plate fairings over cabs indeed lower the drag when a large box trailer follows it. Unfortunately when the tractor travels alone and the driver has forgotten to retract the fairing the drag is far worse. In an ingenious solution Ogle Design incorporate a sleeping compartment in the space created by the raised flap which is sealed behind with a flexible concertina fairing. The Cranfield study also investigated fairings fitted to the front end of the container load itself. Lightweight glass fibre mouldings of quarter-circle or semi-circular cross section are mounted on the front face along top and side edges. This creates an aerodynamic cavity region and the main airflow does not break away at the edges since the strong lateral outflow is suppressed. These devices, known as 'windcheaters', reduced drag by as much as 27% at zero yaw and 15% at 20° yaw. Further improvements result from attention to the details of interference flow around wheel arches, underbody and the gap between trailer and cab. If all these refinements could be incorporated in a practical design then as much as 50% reduction in air drag coefficient could be realised.

Another problem of the heavy goods vehicle is the generation of large sheets of spray in rain. This creates a potential danger for overtaking cars, especially in dark and foggy conditions. Reduction of air resistance and the incorporation of fairings will also lessen rain spray to some extent–but this alone will not be enough. The tyres cause rapidly moving sheets of water droplets as they 'squash' the water lying on the road and this must be dealt with as well as the rain hitting the vehicle directly. The Dunlop Company are now marketing 'Spray Guard' which is fitted into the space between wheel and lorry structure and damps down the air currents and clouds of spray.

In addition to the total aerodynamic forces which hinder the progress of cars and cause them to swerve, the wind flow has other adverse effects. Windscreen wipers tend to lift off the glass at high speeds, although improvements in design have now resulted from special wind- and water-tunnel tests with 1/20th scale models. Rain flow, and particularly snow accretion, depend on a car's shape. Even small ridges have a noticeable effect. Accretion can, however, be minimised by careful detail design as indicated by small-scale wind-tunnel tests (fig. 27), and avoidance is easier than removal by heat or strong scrapers.

The ventilation of buses has been improved by aerodynamic studies and model tests. Inflow and outflow through windows depend on the local external air pressure, and the flow through the interior. Passengers do not like draughts but at the same time air inlets must minimise fume and dust pick-up. Forced ventilation by fans is one solution, though it demands more power and leads to extra complications. By careful aerodynamic research, however, natural ventilation can often suffice.

Early racing cars were all power and gained high speeds on banked circuits. Today the tracks are flat and large side forces are needed for turning. Large tyres help but increased downward load is needed which cannot be provided by increasing weight. Two aerodynamic solutions have been developed for this, the first being the addition of aerofoils mounted at a negative angle above the body at the rear and foreplanes at the front. The extra drag of these is more than compensated for by the increased speeds possible in sharp turns. In another concept Colin Chapman of Lotus used aerodynamic suction under the car body to create download. The undersurface was curved to create a venturi effect which was enhanced further by plastic skirts mounted along the car and which sucked down to touch the surface of the track and prevent a lateral inflow of air.

The aerodynamics of road vehicles will remain a matter of great importance not so much to increase speed but in reducing energy losses and offsetting increasing fuel costs. New fuels may alter the internal aerodynamics of the power units and some novelty may be expected as a result of a massive world-wide effort in researching alternative energy sources. In the 1930s Burt investigated the pneumatic car. In this concept the engine produced compressed air which could be stored in a container prior to transmission to air motors at the wheels. Braking energy could then be returned to the reservoir tank and waste engine heat could be employed to increase the pneumatic air pressure. Was this the ultimate aerodynamic road vehicle? The quest to recover kinetic energy by some regenerative methods during retardation is a great challenge–over 50% of the fuel used by London's buses goes into waste heat in the brakes!

High speed trains

Although the first streamlined trains did not appear before the 1930s the importance of reducing air drag was realised by engineers as long ago as 1855 and, in fact, model wind tunnel tests on trains had been made before the twentieth century. With a still continuing trend towards even higher speeds today measuring aerodynamic effects accurately and designing to avoid their worst consequences has become a major part of railway technology. A typical Intercity train, achieving speeds of 93–106 mph (42–47 m/s), expends 75% of its traction power at the wheels in overcoming air resistance. Already trains capable of 155 mph (69 m/s) are running in Japan (Tokaido) and England (Advanced Passenger Train, APT) and models and experiments are directed elsewhere to even faster running.

Figure 36 shows how the aerodynamic drag of a train is made up and also the kind of improvement that is possible by modern techniques.[21] The 40% reduction comes from several sources, e.g. streamlined nose and tail, a smoother surface with hardly any protruberances like hinges and handles, no gaps between carriages and fewer bogies by placing these between rather than under carriages.

The total drag account of a train however includes other items, viz.,

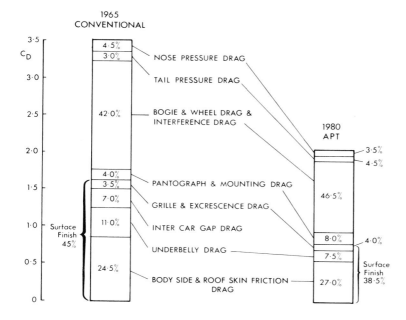

Fig. 36 Breakdown of aerodynamic drag

$$\text{Total resistance to motion} = a + b_1 V_G + b_2 V_A + c V_A^2$$

where V_G is train speed relative to the track;
  V_A is the train speed relative to the air (i.e. with wind);
  a is rolling mechanical resistance;
  b_1 is brake drag and mechanical losses;
  b_2 is air momentum drag created by the energy needed to accelerate outside air into the train for cooling, combustion and air conditioning;
  c is external aerodynamic drag, i.e. $\frac{1}{2}\rho S V^2 C_D$

Energy in accelerating and decelerating high speed trains is conserved by reducing their mass but this increases the ratio of aerodynamic force to weight and thereby increases the possibility of the train being blown off the track. To circumvent this extensive measurements of the side loads created on trains by cross winds have been done in wind tunnels and in full scale experiments. Turbulence depends critically on the ground contours adjacent to the track, high embankments are vulnerable and sharp vortices and eddies can occur in the wakes of buildings or hills. It has been found that the worst overturning risks occur only at a few locations and here the erection of local wind fences can help. It is essential to establish very accurate data on this wind effect so that trains can be slowed down by the right amount in very severe gales. An associated problem encountered in high speed trains is the aerodynamic disturbance they create on the environment. This takes two basic forms.

Firstly there are rapid changes of static pressure as the frontal and rear disturbances pass a given point. For the APT travelling at 155 mph (69 m/s) the peak to peak pressure is 10.4 lb/ft^2 (0.5 kPa) at 13 ft (4 m) distance from the platform edge. This would be doubled if the train nose were blunt shaped. Secondly the train motion induces a slipstream that is carried along with the train and also by the wind. The most severe condition is on the lee side of a train on a windy day when the boundary layer, eddies and wakes of the various parts of the train itself combine with the variability of the strength of the wind. Guidelines are issued, on the basis of considerable aerodynamic research, for safe clearance distances from trains for passengers on platforms and maintenance workers alongside the track. There is also a sharp pressure pulse as two trains pass on adjacent tracks. This can be 30–40% greater than that experienced by an observer at rest.

The momentum drag of air taken into the train has been referred to already. The pressure distribution around the train's surface is very variable with some areas at a high pressure. At others there is a suction tendency and there can also be places of variable pressure, so it is not easy to locate exhausts or cooling intakes for maximum efficiency. An acute problem to be avoided is the ingestion of diesel exhaust gases by air conditioning intakes further down the train. This problem is compounded in particularly long tunnels.

Much early work on train aerodynamics was concerned with smoke removal from the puffing chimneys of coal burning locomotives. Side deflector plates ahead of the smoke box and later wedge shaped false fronts succeeded in raising the smoke clear of the driver's cab windows. The modern need to power trains without oil has increased the use of electric power and this has brought about yet another aerodynamic problem. It is now commonplace to feed electric power down to the locomotive from overhead metallic cables which are at high voltage to save electric losses and are insulated by the surrounding air more cheaply than would be possible from ground rails. Power is collected from the overhead cables by pantographs or sliding bars which operate on a spring controlled linkage to press up and make sliding contact. At high speeds aerodynamic lift forces have been observed to weaken the contact force between pantograph and rail and seriously impair electric current collection. New designs have evolved which are aerodynamically 'neutral' but there is a residual problem of the suspended electric cables themselves which can be blown sideways by a cross wind, locally increased by the airflow over the carriages as they pass beneath. Cables have been known to gallop which exacerbates this phenomenon. Once an overhead electric cable swings out and under the edge of a pantograph arm there is an inconvenient and expensive disruption of several hundred metres of overhead 'knitting'.

There are special aerodynamic effects caused as trains pass into and through long tunnels. Likening the train to a loose fitting piston moving in a cylinder, the aerodynamic cushioning will depend on the blockage ratio (cross section of train divided by that of the tunnel), the speed and the nose shape.

For a high speed train in a 2.5 mile (4 km) long tunnel and a blockage ratio of 0.6 the aerodynamic power (to be overcome in propelling the train) is twenty times greater than it would be in free air. In a more open tunnel of 0.2 blockage this increase would be limited to only 50% more. The power required and the head on pressure pulse between passing trains, which is a distinct inconvenience to passengers, can be avoided by reducing speed to about 109 mph (50 m/s). It is now recognised that these effects with high speed trains are a major factor in the design of new long tunnels such as that proposed for the English Channel (25 miles, 40 km). Apart from the extra expense of large tunnels of low blockage, questions arise of the desirability of two separate tunnels with relieving passages for the air pulses or large tunnels combining two tracks. There is an energy advantage in a tight fitting long tunnel if traffic is dense since one train compressing air in front of it helps the one ahead by increasing pressure behind it. This concept was taken further in a proposal by Battelle Institute of Geneva to propel a train purely by air pressure differences. The train fitted the tunnel almost without leakage and fans mounted outside the tunnel were arranged to increase pressure behind the train and decrease it in front. An even more startling proposal by Foa of the USA embodied a very tight fitting train in a very long tunnel some hundreds of km long which propelled itself by drawing air in at the front and expelling it at the rear after combustion like a jet engine. Such extreme solutions do not seem likely to materialise. The complex interaction between compression and rarefaction waves set up as trains enter and leave long tunnels has been the subject of special theoretical investigations sponsored by British Rail. Only by using large digital computers could the involved aerodynamics be solved and the agreement with experimental results has been quite remarkable. An outcome of this work has been proposals for alterations to the shape of tunnel mouths and exits by providing a flared entry, subtle changes of tunnel shape and ventilated walls between double tracks.

There has undoubtedly been a remarkable revival in railway technology in the last twenty years and as electric train speeds have increased so has aerodynamic research taken on a new significance. It is interesting to speculate on the possible future developments of ground based train systems. In Germany work proceeds on a very high speed train which replaces wheels on rails by magnetic levitation suspension. Speeds of 280 mph (125 m/s) are in view in this work. Japan has gone even further with investigations into trains employing super-cooled cryogenically levitated suspensions with speeds in excess of 435 mph (194 m/s) as a target. It is too early to say that such systems will prove economical but should they do so even more problems of aerodynamic origin will require to be solved.

Alan Wickens, Director of Research of British Rail, has pointed out that the air resistance of a streamlined train in a close fitting tunnel in which most of the air has been evacuated could be reduced to only 1% of that of a single unstreamlined carriage in the open. This provokes the thought of an underground train motivated entirely by vacuum created in the tunnel ahead of it. Cesare Marchetti of IIASA, Austria, has gone a stage further in proposing a Universal Underground railway using such a principle as a total solution to world travel in a future highly populated world which is nevertheless desperately short of energy. There is evidence that underground trains will grow in popularity especially in some of the growing Megacities. Even if they are eventually propelled by vacuum power there will remain aerodynamic problems, not least of which is to allow the passengers to breathe comfortably.

Ships

The air drag of ship superstructures has been measured in several tunnels, for the force may be 20% of the total air and water drag and can be noticeably reduced by streamlining. Typical drag coefficients are: for a very old angular tramp-steamer, 1.2, reasonably large liner, 0.7, ultra-streamlined ship with no detail projections, 0.2. An equally important aerodynamic contribution to ship design is smoke abatement. Ship styles after World War I favoured squat funnels of roughly square side elevation. An undesirable consequence was the down-wash of smoke into the wake of the funnel, particularly in slight cross-winds. Wind-tunnel tests were able to reproduce this effect on models, and several successful solutions were found. The *S.S. United States* had large vortex-generating plates beneath the funnel top, and in *T.E.S. Canberra* the exhaust is taken well clear in high tube-shaped funnels which have a proportionately smaller width of wake. The airflow over a ship superstructure also influences the landing of helicopters and aircraft.

Yachts

Yacht and sail design have evolved over 5000 years by trial and error, without a comprehensive theory of sail aerodynamics being established. The art and design are by no means unprogressive, as new hull forms and new materials lead to faster craft and more efficient sail shapes. Wind-tunnel tests have been made on a variety of sails by official research organisations and several other groups, and there is increasing co-ordination in the research being done by such bodies as Southampton University and the Amateur Yacht Research Society. The crucial features noted in these model wind-tunnel tests are: the surface quality of the sails (for roughness is claimed to have an adverse effect on performance); the wind gradient upwards from the surface; and the complex interaction between sail and sail, and between sails and hull, which determine the stability. It is helpful to analyse the characteristics of the 'static' sail but this separate element is a small part of the whole process of sailing. A great deal more scientific research would be needed, however, before the 'aeronautical' pattern of research could significantly help the traditional method of building followed by trial. In spite of these investigations the sail seems unchallenged, although a revolutionary proposal was advanced in 1951 for a 'Triscaph', in which the conventional canvas was replaced by double biplane rigid aerofoils mounted on a 'Y'-shaped mast.

Environmental power assistance for ships

This term is used for new applications of wind power to commercial ships. It is being actively pursued in several countries, e.g. USA, Sweden, Japan and the UK, in response to the increasing cost of oil fuel and the ultimate prospect of severe shortage of this otherwise almost universal energy source used by world shipping. There are two basic approaches: one is the re-emergence of the nineteenth century fast sail trading ship using new technological advances in aerodynamics, sail materials and structure. Presumably such a ship would have a small auxiliary motor for manoeuvring or overcoming doldrums. The second is the use of various wind energy systems (including modern sails) applied in very large vessels which nevertheless have also a fairly large diesel power source for manoeuvring, but primarily to maintain schedules in protracted periods of light winds.

The new aerodynamic contributions are wide ranging. Sails made of modern synthetic fibres have high strength, low stretch, zero porosity and increased smoothness. Using also aerodynamic techniques learnt from aeronautics, such as leading edge slots and high aspect ratio, it is claimed they provide twice the driving force compared to their last century counterparts. Exploiting the variable and uncertain natural winds is helped by precise world-wide radio/radar/satellite navigation systems and the convenience of satellite weather pictures received on board by tele-facsimile. Masts and rigging can be lighter and smaller with new materials, thereby appreciably reducing air resistance, and there seems some hope of reliable automatic systems for sail handling to minimise crew requirements. Alternative aerodynamic means of capturing wind energy are: towing by means of large kites, horizontal–or preferably–vertical axis wind generators driving electric motors and propellers and a revived Flettner Rotor ship. This was first demonstrated in the 1920s and consists of one or more large vertical circular cylinders mounted above the boat, each driven by a power source. In a wind the rotation of the cylinder creates a lateral force (Magnus force, in chapter 3) of worthwhile proportions. The early demonstration was unsuccessful due to fitting the rotors on unsuitable hulls, which made them subject to capsizing in strong winds. Latest concepts include retractable cylinders (on the principle of the telescope) to avoid this difficulty and augmenting the Magnus flow by compressed air fed to slits let into the surface to increase circulation.

The situation is rather analogous to that of the airship. There are many new possibilities using modern techniques but the research and proving to get a good combination in scaling up to large, effective sizes will be long and difficult. The Japanese have taken an initiative by mounting two masts with large folding hard sails on a 1600 ton (1.63 Gg) trading ship. These are unfolded when the wind speed is favourable and, it is calculated, will save 10% of fuel costs. At present it is not yet economic–but as the price of oil rises further it may well be so in the future.

Hovercraft

The hovercraft,[22] cushioncraft, or ground effect machine, is a new mode of transport whereby the craft rides on a cushion of air whose pressure is increased above normal by fan-supplied energy (fig. 37). It first appeared in

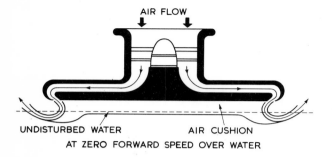

Fig. 37 Airflow in hovercraft

England in 1959 and soon showed its versatility by travelling over land, sea, mud, marsh and ice. The illustration is of the first type that actually operated successfully and the air cushion underneath is maintained by a peripheral jet of compressed air, which creates a curtain of moving air which divides the inner higher pressure air from that outside. The peripheral jet is directed inwards and downwards and is eventually curved outwards by the higher cushion pressure as it makes contact with the ground or water and then flows over the surface away from the craft. This arrangement gives the hovercraft a far greater clearance above the ground, is more stable and enables it to ride over greater undulations than if it merely had a large inverted box arrangement called a plenum chamber. As hovercraft grew in size and speed and had to contend with larger obstacles, rougher seas and greater wave heights the surface clearance was further increased by the addition of rubber skirts. Many designs were tried but the one illustrated in fig. 38 was developed for the British Hovercraft Corporation's Super-4 Hovercraft (fig. 39), which weighs 300 tons (305 Mg), has a cushion area of 11 500 ft^2 (1070 m^2) and cruises at 65 knots (33.5 m/s) over calm water. Air is pumped into the skirt bag which inflates so that it acts as an elastic buffer against severe wave impact, yielding so that it increases the cushion area and gives a greater restoring effect. The air then passes through perforated membranes into the large number of separate 'fingers' which are U-shaped and convey air inwards and downwards into the air cushion yet exposing a curved profile to the wave surfaces. The anti-bounce web restrains the outer skirt from distorting as a trough follows behind a wave and the hovercraft rides above the water surface.

There are many aerodynamic features of hovercraft which can best be subdivided into internal and external. The first includes means of taking in large quantities of air for the cushion and possibly for the engines, filtering it of salt, spray and sand, compressing the air by fans and ducting it to the skirt

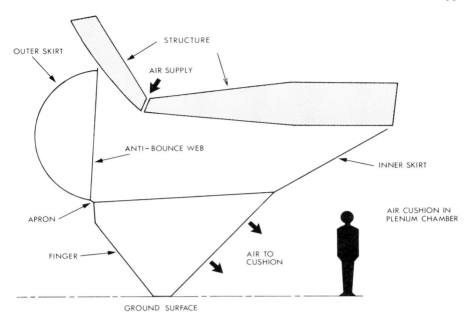

Fig. 38 Rubber skirt of hovercraft

Fig. 39 BHC Super-4 Hovercraft

as previously described. The power to achieve this is termed 'lift power' and varies as the cube of the volume flow measured in cubic metres per second. The external aerodynamics deals with propulsion which is usually provided by air propellers mounted above the body and driven by gas turbines by means of shafts and gears. Directional stability and control is achieved by fins and rudders and/or rotating propellers especially at low speed. The airflow over the craft, past the propellers, control surfaces, skirt, etc. creates the air resistance.

The choice of the cushion pressure and the resilience of the skirts influence the response of the hovercraft to wave impacts. Rapid vertical impacts tend to create disturbing vibrations and particular combinations of wavelength (crest to crest), hovercraft speed and cushion characteristics can set up a resonance effect which is best avoided. To further stabilise the air cushion, which is quite a large volume in the great plenum chamber under the body, two stability bags are installed underneath, one running the length of the craft as a keel and the other at right angles to it. The compartmentation assists in providing righting moments when the craft tilts. Pressure rises in downward going compartments and vice versa.

Clearly the interactions of the internal and external aerodynamics are quite complex and many experimental techniques have been evolved to sort out the separate effects during the design process. One-sixth scale models of the skirt and fingers were tested for inflation shape, resistance to flow and response in simulated wave impacts. These were then compared with the designs generated by complex computer calculations. The overall stability behaviour was examined in a one-twelfth scale free flight dynamic model equipped with petrol engines, variable pitch propellers and operated under radio control. A similar sized model was also towed in a research water tank at different speeds and wave conditions. Previous fast surface marine systems have also encountered complex interference effects between rapidly moving air and the mobile water surface and there is no doubt that the hovercraft could not have made such remarkable progress in the last twenty years without careful use of those advanced techniques.

Ambitious improvements are still being researched under the spur of increasingly high fuel costs. These are now typically 50% of all running costs and are $2\frac{1}{2}$ times as costly as any other single item although maintenance costs are now rising too. BHC have published a comprehensive analysis of the technical developments to attack this problem. Reducing lift power by 50% from its present value of 5000 hp requires attention to filtration losses and filter design, reducing volume flow by more resilient skirt design, improved intake shaping and efficiency, reducing air gaps at the cushion corners and more efficient fan blades. In order to reduce air drag a streamlined overall body shape and two large ducted fans (which also reduces noise instead of four propellers) could contribute to a 35% overall improvement of water and air resistance. Fuel consumption is reduced by these improvements and further still with modern gas turbine units themselves using 25% less fuel than their predecessors. A good index of overall efficiency is the ratio of power to total mass which has fallen from a value of 120–140 hp/ton (88–103 W/kg) in 1960 to 50 hp/ton (37 W/kg) today. A future target of 40 hp/ton (29 W/kg) has been set by BHC designers. Concurrent with these improvements are better reliability of service resulting from the increased ability to operate in stronger winds and rougher seas, largely from better skirt design.

There are other ways of designing hovercraft. In France the Bertin Company creates an air cushion in a series of independent circular 'jupes' or flexible skirts. These have been used on sea-going hovercraft and also applied to lorries giving them a unique amphibious capability. In a sidewall hovercraft the skirts along the sides are replaced by solid thin walls which always penetrate beneath the water surface. Fore and aft cross skirts trap the cushion. Since these craft can never leave the water they conveniently use a water propeller or water jet (which are quieter) but have not proved to be very popular so far.

For the near future 60% fuel savings for the English Channel hovercraft are technically within reach. This would make economic sense and probably permit extended operations to service North Sea oil platforms. Many projects for larger and faster hovercraft have been published. Very large size is needed to cope with large ocean waves and swells and projects upwards of 1000 tons (1 Gg) have been shown by Bell in the USA. Pure ocean hovercraft could use sidewalls but amphibious types would need flexible skirts of enormous depth. Even nuclear powered versions were proposed as a means of collecting Alaskan oil instead of constructing a pipeline. Although increased size does seem a possibility, increasing speed raises other kinds of problem. The dynamic pressure of the external air encountered in front of the leading skirt increases as the square of the speed until a value is reached when it well exceeds the cushion pressure and collapses the skirt. All the external aerodynamic forces increase in a similar way affecting intake design, propeller efficiency and stability. Increasing speeds to 100 knots (51.5 m/s) would be a big technical step but on present evidence hovercraft are unlikely to travel much faster than that. There is already another hybrid on test–the ramwing. This is a winged aircraft, propelled by jet, which flies above and clear of the water but very close to it. It traps a ground effect pressure beneath its specially shaped wings and its speed is more limited by automatic control matters and safety than aerodynamic interactions with the surface. Very large transocean versions have been schemed but it is premature to speculate on the outcome of this novel craft. It has reached a stage of development in 1980 that the hovercraft attained in 1960. Who could say how far it could develop by A.D. 2000?

Before leaving this subject it is worth noting that the flexible cushion developed for the hovercraft has been applied to moving heavy objects on roads which could otherwise not support the concentrated loads of wheels. A flexible bag landing system has been fitted and flown on a test aircraft in the USA dispensing with the normal undercarriage. This sounds potentially very attractive with the prospect of landing on grass or away from runways, but apart from the one successful experiment nothing further has developed.

Creating the cushion pressure quickly enough on landing before the body touches the ground must be a difficult problem to overcome because of the collapsing pressure of the external air, already mentioned as a limiting factor in hovercraft speed.

In all the foregoing groups of transport, it is not easy to measure the full-scale air drag accurately. In cars, trains and ships, some of the engine power is lost in transmission; rolling friction or wave resistance is a large force and there are several interfering effects. The aerodynamic drag, which is obtained as the difference of two large quantities, is thus at the mercy of all the errors. It is only by building up a store of knowledge from many model tests, supported wherever possible by partially representative full-scale tests, that these errors can be kept under practical control.

INDUSTRIAL AERODYNAMICS

Moving air is widely used in industrial and chemical processes. It provides oxygen for combustion in furnaces and in diesel and petrol engines. Pneumatic systems use air as a medium for transmission of energy. Air is used also to transfer heat to and from machines to maintain desired temperatures. A special feature of industrial aerodynamics, which distinguishes it from transport aerodynamics and aeronautics, is that the air is usually forced to flow in pipes, ducts or passages. In many instances the boundary layer may extend over the whole flow field and thus exclude a main flow as described in chapter 3. (In heat engines, where air is a medium for transferring heat and mechanical energy, most of the changes in the condition of the air are adequately described by thermodynamics, and aerodynamic motions do not often have to be considered in detail.)

Steelmaking furnaces

Large quantities of gas are blown into blast furnaces to melt steel from ore, and a typical flow is 50 000 ft^3/min (23.6 m^3/s) at 15 lb/in^2 (103 kPa). It is clearly difficult to observe and measure the flow, which has wide temperature variations, different chemical compositions and very complex flow patterns. Scale model tests have been made of furnace flows at 1/12th size and with the correct Reynolds number. In some parts of the furnace where the gas velocity is very low, less than about 15 ft/s (4.6 m/s) and at a high temperature, say 1500°C, no instrument can measure the air speed at such high viscosity and low dynamic pressure. At the other extreme, oil-fired furnace flames reach 1000 ft/s (305 m/s). The radioactive gas Radon has been used as a tracer of blast furnace gas with minute concentrations of 1.2×10^{-15} cc/cc of air.

Aerodynamic research techniques have also been applied to the design of open hearth furnaces since the 1950s. The basic aim is to increase efficiency of combustion and minimise erosion of the refractory brick linings. The former is assisted by a steady flow and the latter is caused by vortex flow and the transport of high temperature oxide particles. The work of Chesters[23] affords an excellent example of the retrospective application of aerodynamic knowledge to an established engineering practice.

The refractory furnace bricks should not normally exceed 1650°C and yet the hot-flue gases are 1700°C, so clearly the margins are small. Research was concentrated on small aerodynamic models, accurate water-flow models and means of flow visualisation such as aluminium particles and air bubbles were explored. Heating effects were investigated by 1/12th scale hot models with refractory linings, and it was shown that buoyancy effects are small mainly because combustion is progressive as the air passes along the furnaces. The Reynolds number is generally about 100 000. Furnaces use either producer gas, or oil-fired air in which the aerodynamic processes are somewhat different. An indication of the severity of the combustion is that the heat-release in the flame is of the order of 100 000 Btu/ft^3 (3.73 GJ/m^3). (The heat-release in a jet engine is fifty times greater, but the metal combustion chamber walls are protected as the flame is diluted with 100:1 excess ambient air.) It was shown that rotary flows having a marked effect on performance are very readily induced in typical furnace shapes, and basic research was made into the flow of jets into such cavities (fig. 40).

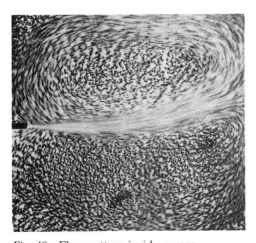

Fig. 40 Flow pattern inside square

Notable results of this work were the redesign of air passages leading to and from the furnaces, weakening of vortex flow in exits which were the cause of excessive brick erosion, and general improvements in combustion efficiency by attention to the detailed shape of the furnace combustion space (fig. 41). This work led to discussion with other aerodynamicists and the evolution of a new shape of furnace. Hence there were significant cost reductions in the repair and replacement of refractories: savings in refractory costs were 7½p per ton of steel made, and roof repairs were reduced by nearly two-thirds.

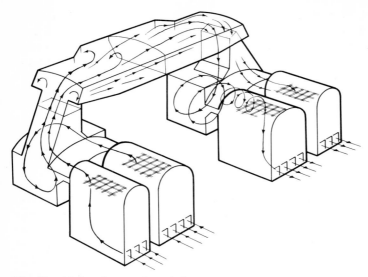

Fig. 41 Airflow in open hearth furnace

Although such improvements justify this type of research and confirm its accuracy, the aerodynamicist is fundamentally more interested in the correlation between model and full-scale airflow. In spite of the general difficulty of full-scale measurement, checks have been made, e.g. by introducing wood tracers whose surface is burnt in the direction of airflow, by observation of flame colours, and filming through small glass ports. The general impression is that small, relatively cheap model tests are reliable and useful. This sort of aerodynamic research into steelmaking has stimulated similar work on brick-kilns, flame research in boilers, airflow inside large gas works and steel shops, and even the production of more efficient cremation furnaces.

Apart from the essential importance of evolving efficient furnace equipment, the introduction to steelmaking of automation has also made necessary the measurement of blast furnace conditions for control purposes. Radioactive tracers have been used for this, and now television and water-cooled cameras are also being used.

Heating and cooling by air

Air is widely used as a medium for transferring heat. The electric hair-dryer is an example of air used for heating, and the familiar large concrete cooling tower of the electric power station is an instance of convected air-cooling by water. Both natural convection heat transfer and forced airflow-cooling are methods commonly in use.

Large electrical transformers produce waste heat which must be removed continuously to prevent excessive rise in temperature. In a conventional transformer the windings are immersed in oil which rises on being warmed and then circulates down through external cooling pipes which are exposed to the natural wind. This design provides a means of cooling but has in fact been found inadequate. The velocity of oil in the pipes is slow and the temperature rise of the external pipe surface too slight to afford much cooling. Transformers are now often cooled by air blown by a fan through ducts surrounding the windings. Although the air fans require electric power, the overall effectiveness is greater.

Air conditioning of buildings

Ours is a civilisation notable for the heating and ventilation of its public and private buildings. The important factors in either case are the maximum quantity of heat, the distribution of heated or cooled air by fans and ducts, and the degree of mixing. There are thus several problems, and as such systems are costly experiments are used to arrive at good design principles. Perhaps the most fascinating ventilation and heating tests were made for the Debating Chamber of the rebuilt House of Commons in 1950. In a quarter-scale model at the NPL, cooled ventilating air was introduced into ducts high in the walls so as to descend and mix with the Chamber air, without (it was hoped) undue draughts and cold spots. The heat output of each Member was represented by a shielded electric lamp (the convection of heat from 600 Members is not inconsiderable), and measurements of duct velocity, temperature and chamber flow patterns were made at a distance, so that no human observer should distort the airflow. After preliminary tests had shown that Reynolds number effects were not overwhelming, the velocities were reduced by a quarter of full scale, and the temperature gradients increased fourfold. The improvements resulting from these tests included halving the designed ventilation supply, and reducing draughts to an acceptable level of comfort.

Ventilation of mines and convective flow in refrigerated food stores have also benefited from aerodynamic experimental techniques.

The vortex tube

This is a remarkable aerodynamic device which can produce a flow of cooling air lower than the supply temperature by inducing a vortex flow into a two-way pipe through tangential inlet slots (fig. 42). The cross-flow induced by the vortex creates a cooler inner core, which leaves by one exit, and a warmer outer tube of air which leaves by the other. The exact explanation of this phenomenon is still uncertain, but experiments have been made to determine good shapes and performance.

The temperature drop obtainable in practice is limited to 30–40°C, but the device is attractively simple and has no moving parts. It has been applied experimentally to aircraft and missiles but has not become established, probably because it has so far been approximately a hundred times less efficient than the conventional liquid evaporative refrigerator, and is extremely noisy. It has, however, been used as a temperature-measuring device for aircraft.

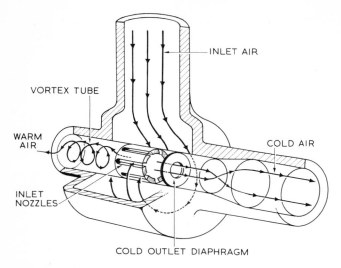

Fig. 42 Flow in vortex tube

The Vortec Corporation of Ohio, United States of America, has persisted with the development of the vortex tube for a variety of applications where a continuous mobile supply of cold air is required in local areas, which are often awkward to reach by any other means. Whilst the refrigerator is more efficient and virtually noise-free in reducing temperatures within a closed space the Vortec Cold Air Gun directs a small cold jet in which the moving air carries away the unwanted heat. It is quite small–1.5 in (38 mm) in diameter and 7.8 in (198 mm) long. Applications include cooling hot spots in manufacturing processes, rapid solidification of molten solder in constructing electronic circuits, and cooling tool and workpiece during drilling, grinding and milling operations. Better tool life and less overheating of the work results.

The gun is supplied by ordinary industrial compressed air–80–100 lb/in^2 (550–690 kPa)–and ingeniously employs the hot air generated by the vortex to eliminate condensed moisture which might otherwise interrupt the air flow by ice blockages.

Sprays and atomisation

The distribution of liquid in the form of drops in a moving stream of air occurs in many industrial processes. In the diesel engine, oil is sprayed into the compressed air in the cylinder at pressures of about 1000 lb/in^2 (6.9 MPa) in about 1/250th sec. The droplets, about 3.3×10^{-6} ft (50 μ) in diameter (μ = micron = 10^{-6} metre), must ignite rapidly as the whole explosion stroke only lasts about 0.02 s. A less rapid process occurs in the oil firing of furnaces for ships and steam generating plants. In the gas turbine a very fine atomisation is required into the high velocity air from the compressor. In many of these processes the atmosphere plays a relatively small part, and the degree of atomisation depends on the type of atomiser (i.e. hydraulic injection, swirl chamber atomiser, air blast, or centrifugal disc) and the physical characteristics of the liquid, particularly its surface tension. In other processes the interaction between the liquid droplets and the surrounding air is all-important (fig. 11). There are roughly four patterns in which the liquid breaks down into droplets (a study in itself), but generally the liquid emerges from the nozzle as a sheet which is progressively torn apart by the airflow and surface tension to form a variety of drops. Normal sizes range from 50–100μm diameter, but in disinfectant and agricultural spraying droplets as small as 10μm are required.

The time taken by a droplet to acquire the speed of the atomising airstream is very short. The equation of motion is:

$$\frac{\pi}{6} d^3 \rho \frac{dv}{dt} = \frac{\pi}{8} d^2 C_D \rho_0 (u - v)^2$$

where v = droplet speed, u = airspeed, ρ = liquid density, ρ_0 = air density, d = diameter, t = time, C_D = drag coefficient. The time taken by droplets of different size to achieve 90% of the airspeed is shown in Table 10.

Table 10

Diameter (μm)	Time (s)
5	9×10^{-5}
10	3×10^{-4}
15	5×10^{-4}

Aerodynamic capture of particles

The accuracy of very precise electromechanical instruments is sometimes impaired by contact with tiny airborne particles of dirt and dust, and a new manufacturing technique has had to be developed to protect such mechanisms during manufacture. The largest particle tolerable, for example, in the guiding instruments of a guided missile is about 7μm in size and for gyroscopes, 4μm. Small particles are formed by being broken off the sharp edges of metal parts and are produced by locking washers and knurling, and there is, moreover, a general transport of fine iron particles and dust from clothing, etc. in the atmosphere. To prevent contamination clean-air laboratories are used where such foreign bodies are eliminated by filtration and special protective clothing is worn by operators. The airflow through the conditioning systems and over work benches in the laboratory has been the subject of study. For example, by keeping the laboratory air at a small positive pressure and using air locks, particles are prevented from entering when the door is open. Small equipment manipulation is done in pressurised clean-air cabinets into which the operators pass their hands and wrists through air-tight ports.

Fluidics

Industrial use of pneumatic (pneuma = Greek for wind) power is well known: air under pressure restrained in pipes and vessels does work on pistons, valves, pumps and turbines. But from about 1960 a new technology has arisen in which air (or more generally fluids) is used for control purposes by measuring, computing and amplifying as part of automatic systems.[24] Many complex titles were originally given to this new technique such as 'pneumatic automation', 'hydraulic logic', 'fluid state systems', 'fluidonics' (by association with the work 'electronics' which performs similar functions but uses electrons instead of fluid molecules) and lastly 'fluidics', which seems to have survived as its lasting title.

The general principle employed is to combine two or more discrete streams of air in such a way that the presence of the second (or subsequent) flows alters the direction of the first. Also a small control flow can change the direction of a much more powerful flow. The first historic example known was created by Ludwig Prandtl, the great German aerodynamicist, in 1904. As shown in fig. 43(a) a lateral flow of air introduced into the bell mouth of a diffuser will persuade the flow to adhere to one surface, or if two are applied force the flow to expand in contact with the conical surface, which it will not do on its own. In 1916 the Italian scientist Tesla, well known for his work on high frequency electricity, patented his 'valvular conduit' seen in fig. 43(b). The re-entrant passages permit fluid to pass freely in one direction but encounter high resistance when the flow is reversed. Such an effect is analogous to that of a thermionic electric valve or diode, and can be used similarly to amplify fluid flows.

Firstly some basic fluidic devices will be described. An *amplifier* is an element which employs a low energy control signal to change the power output of the device. A *sensor* is sensitive to changes in operating conditions such as pressure or temperature and its output varies in response to these changes. A *transducer* is a device which converts signals from one energy form to another, e.g. an electrically driven diaphragm which creates a fluid pulse. An *oscillator* produces an oscillating fluid output from a steady input. Devices are constructed from several kinds of fluidic principle, e.g. wall-attachment, axisymmetric focussed-jet or turbulence amplifier; to illustrate the operation of several devices only wall-attachment arrangements will be treated.

The fundamental method is shown in fig. 43(c). The main jet, emerging from a sharp edged orifice creates a low pressure vortex bubble between the outer edge of the jet and the expanding wall. The greater fluid pressure in the jet curves it towards the surface to which it sticks in a very stable manner. Injection of a lateral flow at the lower port guides the jet flow across to the

Fig. 43 Fluidic devices: (a) Prandtl diffuser; (b) Tesla's diode; (c) wall-attachment element; (d) oscillator; (e) AND gate; (f) OR-NOR gate; (g) half adder; (h) vortex device

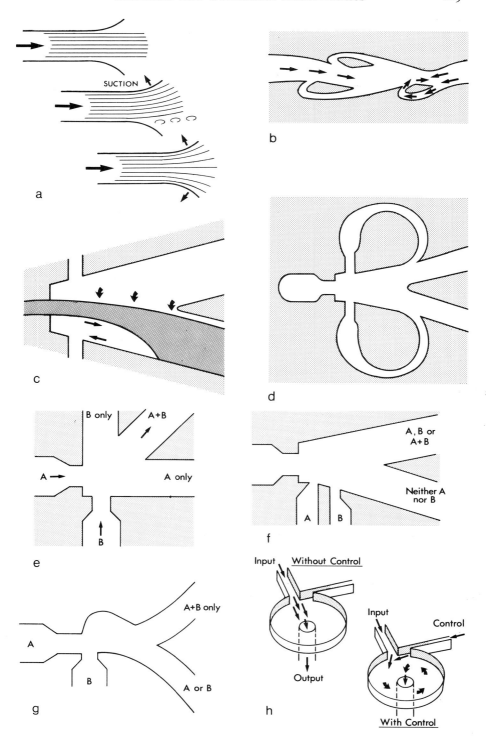

opposite wall and exit to which it again persists in a stable manner. This is termed a 'bistable switch'.

This phenomenon of a fluid jet flowing in contact with a solid surface and being constrained to follow it even if it curves away was discovered by accident by the Romanian engineer Henri Coanda in 1910. He had built a revolutionary aircraft propelled by a kind of jet engine. An air compressor driven by a reciprocating engine fed compressed air to a burner where fuel ignited and produced a fiery jet exhaust. Alarmed about the possibility of igniting the plywood and fabric fuselage Coanda fitted metal plates alongside the exhaust to shield it from the wood. Unfortunately the jet was sucked towards the side plates and his preoccupation with the resulting fire led to a serious but luckily not fatal crash. Coanda later discussed this unexpected aerodynamic phenomenon with Von Kármán, the great Hungarian aerodynamicist, who named it the Coanda Effect. It has been applied in many ways; to nozzles, wings, ground effect machines and hydrofoils. Coanda could hardly have imagined it being used in minute devices embodied in automatic control systems. Much of aerodynamic knowledge is discovered accidentally as the result of creative and observant people trying out new concepts. What may not appear to be particularly useful at first may often be applied in a revolutionary manner many years later.

The basic wall-attachment device can be turned into an oscillator by diverting a small part of the jet from the divergent wall and turning it back to become the lateral control jet (fig. 43(d)). The delay as the newly established jet travels around the loop determines the frequency of oscillation.

For computing and decision making an automatic system requires logic elements which combine two signals in special ways. There are six important types the simplest of which, the switch, has already been described. This (sometimes called a bistable element) can be adapted to respond to the difference between two control streams, and the power input jet will move away from the larger of the two control streams. A four-input bistable element with two control inputs per side will respond to the difference between the sums of the two streams on each side.

Taking the same nomenclature as used in electronic computing circuits fluidic elements can represent AND, and OR-NOR gates as in fig. 43(e) and (f). AND only provides an output at A when both streams are present. The OR power jet moves to the upper exit limb if either or both control streams are present. Figure 43(g) is the half adder which gives one exit state for either power input and another if they both occur simultaneously. By combining such fluidic elements complex computing and control systems are constructed. Vortex devices, fig. 43(h), employ a different method permitting flows to combine in varying ways and also providing markedly different resistances when a flow is reversed in direction (as the Tesla valve does).

Turbulence amplifier

A laminar flow power jet enters a vented cavity and, when no control input is present, leaves by a tube in the opposite wall aligned exactly with the incoming flow. Any control signal entering the cavity space laterally to the control flow makes this turbulent and very little of the entering stream reaches the exit. Very small flows can switch the main jet. Because of the need to maintain laminar flow however it is sensitive to Reynolds number.

Fluidic elements can be so small that as many as ten can be arranged within a cubic centimetre. The air velocity should not exceed the speed of sound and the Reynolds number must exceed a value between 2500 and 3000 to permit the flow to be turbulent and able to adhere to the surfaces. At lower Reynolds number a laminar boundary layer persists which can make the jet unstable and hence unreliable. Throats are 0.006–0.02 in (0.15–0.5 mm) wide and the time for fluid to flow across an element (transport time) can be as low as 1 μs increasing to tens of milliseconds for larger units. Response (switching) time is from ten to twenty times greater than this. Transport time is lowest for hydrogen, other gases being slower by factors of 3 to 5, oil by 50 times and mercury by 200. Amplification ratios up to 10 are achieved by a single element and when ganged together as much as 1500 could be attained in a large system. Fluidic oscillators can have a frequency as high as 100 000 Hz; within combined systems this may fall to only 2.5 kHz.

In hybrid systems pure fluidic devices may be combined with moving balls or spools which can be displaced by fluid flow either to direct piped flow into different passages or completely seal off an input flow. The latter is desirable in some instances to reduce power consumption, but since the fluid has also to accelerate the mass of the moving part its response is far more sluggish.

Applications seen for fluidics in the 1960s were very wide, ranging from a fluidic toothbrush actuated by water from the bathroom tap to windshield wipers, car power steering systems and office computers. One of the more successful uses is for thrust vector control on spacecraft or guided missile rockets. For propulsion a high speed jet of combustion gases is ejected aft from the nozzle, the thrust being aligned with the axis of the vehicle and the efflux flow being axisymmetric. For control purposes, i.e. to create a sideways force to bring the spacecraft back on course after a disturbance the rocket jet can be deflected sideways by injecting a flow of control gas into the expansion cone of the nozzle. This causes the flow to break away from the nozzle surface downstream of the fluid injection and the main jet thrust is then forced away at an angle so creating the lateral force. Removal of the control gas flow immediately restores the symmetric flow. The Titan space rocket vehicle uses this principle with a ring of control gas ports to deflect the jet in any direction as determined by the guidance system on board.

The new impetus for fluidics in 1959–60 was the unacceptable unreliability of electronic systems in guided missiles. Intense vibration often combined with high temperature damaged the miniature, somewhat fragile, components or fractured the solder in wire joints. By comparison fluidic systems were robust, had no moving parts and could be manufactured without high precision. By 1960 rapid growth of fluidic applications was being predicted with multi-million dollar business expected by the end of the decade. But this did not happen. Instead new micro-miniaturisation techniques for integrated electronic circuits satisfactorily overcame the reliability problem and the rising demand for computing complexity and power, such as that required for the Apollo Moon program, could only be met by the incredibly minute electronic circuits. But although the fluidic miracle did not occur, a great deal was learnt in the early 1960s, a few guided weapons were flown under the control of fluidic systems and there is a continuing residue of applications in special areas. These include control of corrosive liquids or sewage, slurries and systems in high temperature and radioactive environments. Hence, although most applications are not strictly aerodynamic but are rather fluid dynamic, nevertheless the methods and components can be used with air and remain a most remarkable and inventive incorporation of some special aerodynamic principles.

Aerogenerators or wind energy conversion systems

Man-made machines to harness wind energy can be traced back to 200 B.C. in Persia; they were used extensively in Europe during the Middle Ages and are now re-emerging as a response to the expected world energy shortfall. Aerodynamic aspects are concerned with assessing the wind energy distribution at favourable sites and in the efficiency of conversion offered by the many different designs of aerogenerators now available. As in the past there are several modern applications but the primary emphasis today is in generating electricity.

The Persian windmills for grinding grain were vertical axis panemones, a type still in use today in China for pumping water. In this type sails hang from radial poles mounted above a vertical axis and their angle can be changed during rotation by ropes or linkage to expose their full surface when going downwind and to present a minimum area when moving upwind. The Persian panemone was simplified by enclosing it in a tall building which shielded the prevailing wind from one half of the blades and speeded it up past the other. The more conventional horizontal axis windmill was probably first used in S.E. Europe using fabric sails stretched over rotating poles mounted on fixed towers. The English began constructing windmills from 1190 and the Dutch developed them extensively for draining marshland and flood control. They invented the tower mill in which the upper cap can turn into wind. The horizontal axis propeller type with four, five, six or eight wooden blades was first recorded in the fourteenth century and became the classic design used throughout Europe until the 1930s. Apart from grain milling they were used to pump water, mill oil and paper, and power saws. Aerodynamic control was introduced by the fantail of Lee (1745), which automatically turns the windmill to head into the wind, and the spring sail of Meikle (1772) which spilled wind out of the sails when its force exceeded the preset spring tension. John Smeaton measured windmill sail effectiveness by means of models mounted on a whirling arm somewhat akin to that used by Robins (p. 5).

The whirling arm was turned at a constant speed by hand and as the windmill was rotated by the relative wind it encountered it raised a weight by a cable passing over a series of pulleys. The air speed, windmill rotation speed and aerodynamic driving force were measured and from the corrected results Smeaton deduced an overall efficiency of 28% and determined optimum angles of incidence for the sailboard varying from hub to tip. In 1759 he presented his results to the Royal Society. This must surely be one of the first examples of aerodynamic model experiments ever conducted.

In assessing the power that can be extracted from the wind it must be recognised that the wind is a relatively weak source of energy and large machines are required to collect worthwhile amounts. Typical theoretical values are:

Table 7

| Wind speed | Power in kW–circular areas of diameter (1 kW = 1.34 hp) | | |
mph (m/s)	12.5 ft (3.8 m)	50 ft (15.2 m)	200 ft (61.0 m)
10 (4.47)	0.38	6	96
30 (13.4)	10.4	166	2664
60 (26.8)	83.2	1331	21 300

These theoretical values are calculated from a formula derived by Betz in 1927 from an adaptation of the Froude theory of ships' propellers (page 138). The maximum value which can be extracted from the wind occurs when the final wind velocity (in the wake) has been reduced to $\frac{1}{3}$ of its free stream value. This works out to be 16/27 or 59.3% of the power in the wind, i.e. $\frac{1}{2}\rho V^3 A$ where A is the disc area of the windmill. In practice the airflow is not uniform as the theory assumes and the blades/sails suffer air friction and vortex tip-losses, so that the factor 59.3% is reduced to approximately 35%. Other theories have been advocated such as that of Sabinin which takes into account an additional momentum in the wake induced by large eddies which form at the boundary with the main flow. His theoretical factor is 68.7%. Other values of 63% and 56% have been argued. This may seem an academic point but, as will be seen later, there are many difficulties in measuring the wind energy collected by a windmill arising from the wind's unsteady characteristics and the different responses by wind machines to these changes. Although such speed variations can be eliminated in a wind tunnel this method unfortunately introduces model scale errors such as Reynolds number effects on the boundary layers over the blades, drag and interference and also the difficulty of representing the steady growth of wind speed above the ground surface.

In their heyday, before the coming of coal and steam power, Europe had 10 000 windmills generally of the classic Dutch design capable of outputs of from 34–50 hp (25–40 kW). In this century wind generators were generally of three classes, viz. small water pumps of 1 hp (0.746 kW), domestic electric generators of about 1.34 hp (1 kW) and some large experimental electric generators. Typical of the last was the 1675 hp (1.25 MW) Smith Putnam machine built in the USA during the 1940s. It employed a two-bladed propeller type rotor of 175 ft (53.3 m) diameter and operated for five years. Somewhat similar systems giving outputs of 134–268 hp (100–200 kW) were built in UK, Russia and Denmark. The best example of the small electric machine is the American Jacobs of which tens of thousands were built between 1930 and 1960. They were reliable with low maintenance cost (less than $10/year) and yet were phased out thereafter by cheap grid electricity. This clearly demonstrates the sensitivity of energy costs and the difficulty of predicting the viability of new systems in the future. Since the 1973 energy imbroglio many of the Jacobs machines have in fact been recommissioned. We are now witnessing a major revival of wind machines as a result of the marked increase of energy prices and the desire of many countries to reduce their dependence on overseas oil supplies. Leaders in this field are the USA, Sweden, Denmark and Japan. The remainder of this section will concentrate on recent experimental work on new concepts, the magnitude of the possible wind energy contribution to national energy needs and the scope of some of the new constructional programmes.

It must be admitted that the major progress is coming from the horizontal axis wind energy conversion system (WECS) now employing advanced techniques of aerodynamics, structures and control learnt from helicopter experience. Powers up to 6700 hp (5 MW) are typical. Nevertheless the alternative designs are of interest, some raising quite special aerodynamic issues. Designs may be classified by vertical or horizontal axis, and whether the wind energy is derived from lift or drag forces. Still others contain the airflow in various surfaces and a few employ electrical concepts. An interesting aerodynamic variant of the classical horizontal axis blade wind generator was the Enfield-Andreau type (fig. 44). The two blades (40 ft (12 m) radius) were

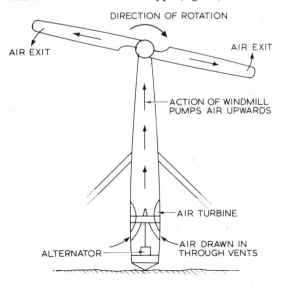

Fig. 44 Andreau 'aerodynamic' aerogenerator

hollow and open at the tips so that as they rotated air was drawn out through the hub and expelled at the tips by centrifugal force. An airflow was created upwards through an 85 ft (26 m) high shaft-like tower supporting the hub and this air current drove a vertical axis electrical generator at its base. This unique aerogenerator was designed for 134 hp (100 kW) output and had the double advantage of machinery at ground level and the avoidance of gearing from rotor to generator. Unfortunately the overall efficiency of about 25–30% compares unfavourably with the 35% of present day conventional WECS. There are four sources of aerodynamic losses, viz. the efficiency of the rotor, internal efficiency as a suction pump (friction losses within the blades and tower), the efficiency of the air discharge and the air turbine efficiency. Another new class of WECS is the Darrieus type (Sandia and Alcon, USA, and NRC, Canada) which also houses the generator at ground level by employing multiple curved aerofoils amounted about a vertical axis. These can operate in any wind direction but are not self-starting; once running the blades rotate with a blade velocity 4–7 times that of the wind speed. This type is the subject of a considerable amount of development in Canada, Sweden and the USA. Drag can be employed to extract wind energy by refined versions of the old Persian or Chinese panemone now called the Savonius rotor. The simplest example of this is the rotating sign at garages (gas stations) where a circular plate, S-shaped in plan, is free to rotate about a vertical axis. Many variations of this simple form are in use–even constructed out of a vertically split oil barrel!

A simple but seemingly unlikely method employs several oblique vertical aerofoils mounted vertically on a wheeled truck transversing a duo rail circuit. The wind pressure creates lift and drag forces (as in a yacht sail) which impels the vehicle along its track and the axle rotation produces electric power by means of a generator. The Free-Wing Turbine Corporation of Salt Lake City, USA, is developing such a system of vast proportions, e.g. 300 ft (91 m) high sails, a $1\frac{1}{4}$ mile (2 km) long circuit with electric power output of 250 000 hp (186 MW)!

Finally three novel aerodynamic forms of WECS show the diversity of means now under investigation and the difficulty of identifying which system will ultimately prove itself to be the classical wind generator of the twenty-first century.

(1) The Diffuser-Augmented Wind Turbine (Grumman, USA) surrounds the rotor by an annular ring which extends downwind expanding into a diffuser cone. The object is to reduce the back pressure thereby increasing the velocity through the rotor and the pressure drop. Blade tip vortex losses are also reduced. Much ingenuity has been used by the introduction of peripheral slots around the diffuser surface allowing high speed wind flow to speed up the boundary layer and prevent separation. Augmentation ratios between two and four are claimed and the system would be suited to very high powers. However, the diffuser is very large, i.e. 295 ft (90 m) diameter and 98 ft (30 m) long, and it may not be economic to arrange for this big installation to turn to face the wind.

(2) More uncertainty centres on proposals to employ vortices to augment wind energy. Aircraft wakes contain wing tip vortices where local velocity is high and pressure is low. Much experimental work of great ingenuity has been completed under the Vortex Augmentor Concept in the USA. Vortices are generated from long slender wing shapes like those of Concorde set at an angle of incidence to the wind. Turbines immersed in the large vortices created by the leading edges seek to take advantage of the concentration of kinetic energy in the vortex. However the turbine interferes with the vortex and may weaken or destroy it.

(3) Another use of vortex flow is the tornado-type WECS of Yen as developed by Grumman in the USA. A large cylindrical tower with oblique adjustable slots along its generators induces the natural wind to create a vortex rising in the tower and leaving through the open top. The upflow entering via augmentor surfaces then drives a vertical axis turbine as in the Enfield–Andreau scheme. The aerodynamics of this are most complex and it cannot be pretended that the vast energy-capturing mechanisms present in a natural tornado actually occur within the confines of the tower. Its power could be enhanced by heating by low grade fuel. Nevertheless, bearing in mind the unique characteristics of the vortex tube there is scope for the discovery of new and beneficial ways of harnessing wind energy.

So much for the machines. What wind energy is available and how much could reasonably be tapped by realistic systems? It has been estimated that the total wind energy of the world is 1.34×10^{17} hp [10^{11} gigawatts (GW)]. [The UK electrical generating capacity in 1980 was 74×10^6 hp (55 GW).] Of this, 10^5 GW is potentially available at the surface of the USA. This represents thirty-times the total power consumption and one hundred-times the electric power production of that country. But where is the wind speed sufficient to make WECS economically viable? Extensive wind surveys are now available for many regions. Annual isovent charts plot places having the same mean annual wind speeds, the five most favoured areas are New Zealand, Eire, NW of the UK, NE Canada and Alaska. At any potential site annual average velocity distribution curves show the speed–time relationship–a first indication of the possibilities for successful wind power use. Actual wind powers developed turn out to exceed those calculated from averaged wind speeds and typically range from 0.003 hp/ft^2 (26 W/m^2) in California to 0.074 hp/ft^2 (592 W/m^2) in Alaska.

The amount of actual WECS construction now in hand worldwide is most impressive. For example, in the USA a 5360 hp (4 MW), 255 ft (78 m) diameter two-bladed horizontal axis machine costing $6 million is to commence operations in 1981. Annually it should supply 10 million kWh (36 TJ) of energy to the federal grid at a cost of 4c/kWh (1 c/MJ). Its blades are of glass fibre developed in conjunction with the Swedish State Shipyard Co. (Svenska Varv AB). These are preferred to steel blades which are prone to cracking. Another project in Wyoming will build forty-one WECS in a group spaced fifteen rotor diameters apart to generate 221 000 hp (165 MW). Another WECS 'farm' at Goldendale, Washington, will operate three Boeing

MoD-2 aerogenerators for a total of 10 000 hp (7.5 MW). Many other US companies are building large units as previously indicated. In Sweden interest centres on modified Darrieus types mounted offshore to gain benefit from the extra $2\frac{1}{4}$ mph (1 m/s) average wind speed which increases energy by about 35%. Consortia in France, Germany and Japan are also active. In the UK Aircraft Design (Bembridge) Ltd., is building an 82 ft (25 m) vertical axis WECS prototype based on a design of P. J. Musgrove of Reading University.

Minor practical difficulties are being discovered with these radically new systems. The 2700 hp (2 MW) two-bladed MoD-1 turbine at Boone, N.C. created infrasound vibrations propagated at less than 5 Hz which disturbed residents by rattling windows and crockery. Such noise emanates from vortex wakes from blades and towers and structural vibration of the blades themselves. Glass fibre blades are expected to damp out much of this environmental nuisance. In UK one WECS encountered difficulties when the pressure oscillations created by the helical vortex wake broke window panes in the greenhouse that was being heated by the power output of the WECS. Another worry is the damage caused by large chunks of ice dislodged from the blades in severe icing conditions. Although none of these is fundamentally incurable–safe distances seem easy to achieve once the problems are quantified–but they may attract media attention and slow down some installations. Legal problems could arise from conflicts between WECS location and forestry, agriculture and wild life. There have been certain objections on aesthetical grounds and yet Dutch windmills were eventually seen to confer beauty and charm to a landscape.

If the US programme goes ahead as planned, by 2000 there should be large WECS available giving 80 million hp (60 GW) electric output (i.e. greater than the whole of the UK in 1980). The capital cost of $75 billion would not begin to return a profit until the year 2003. Apart from the extensive major federal programme a very large number of smaller units for households, farms and factories would be expected.

The extent of the present world-wide revival of aerogenerators is considerable. A great deal of research and industrial effort must be expected to be devoted to the subject over the next twenty years. Considerable aerodynamic knowledge will be required to assist the development of really cost-effective environmentally acceptable systems of high output. It is not beyond the realms of possibility that the eventual winner of the contest has yet to be discovered.

7
Aeronautics

Who layeth the beams of his chambers in the waters:
and maketh the clouds his chariot, and walketh upon the
wings of the wind

Prayer Book

In this chapter the aerodynamics of objects which move through the air without contact with the ground is considered. This category includes insects and birds, airships, aeroplanes, helicopters, missiles and bullets, the large intercontinental rockets and even the golf ball and the boomerang. All such objects can be divided into two classes: 'ballistic' and 'aerodynamic' (see fig. 68). The ballistic bodies, e.g. bullets, shells and rockets, are of simpler shape as they are symmetrical, and their motion is determined by their speed, drag and weight. They therefore travel in curved paths and usually at supersonic speeds which are not often met in nature, meteors being an exception. Some guided missiles have a more complicated type of aerodynamics because they have wings to develop lift forces for manoeuvring, and control surfaces to change the angle of the missile to the direction of the air. In the manned aeroplane, the air is used not only to provide the lifting force, but as the source of oxygen for the propelling engines, and of cabin air for the occupants to breathe. Aeroplanes are not easily made efficient, or safe for that matter, and consequently a great deal has had to be learned about the motion of the air in relation to these different tasks.

Animal flight

Three quarters of the present and extinct species of the world's animals fly but the means they adopt are extremely varied, very enterprising and in almost all respects totally different from those adopted for the aeroplane. A proper study of animal flight[25] would encompass several large books and is inherently complex since observation and measurement are almost always difficult, the deflecting wing surfaces describe unexpected and involved motions and most of the aerodynamics is non-steady. In aircraft unsteady flow is avoided wherever possible since it leads to vibration, inefficiency and loss of control but in nature the subtle and intricate interplay between feathers, wings and unsteady aerodynamics has enabled a wonderful variety of species with very special characteristics to evolve. The main sub-divisions are birds, bats and insects.

It was Otto Lilienthal, the famous German pioneer of hang gliding, who in 1889 very well described the wing motions of a stork in normal flight. A bird's wing consists of two parts, the inner portion operated by the shoulder joint and the outer by a 'wrist' halfway out along the wing. As the wing beats downward the leading edge twists increasingly downward moving outward from the root to the top. In this way it develops both an upward force to counter the weight and a forward component of air force exerted over the upper curved surface of the forward part of the wing. On the return (upward) stroke the wing camber changes to offer the least resistance to the air. This is very evident for example in the extreme case of an osprey (sea eagle) when it makes very large wing motions in order to fly up from the water from a standstill carrying the weight of a heavy fish. So the flapping wings perform the two functions of lifting and propulsion. Wings are jointed at several places and are almost infinitely adjustable for different regimes of flight. In slow steady gliding the outer parts can be swept forward, at medium speeds they are either straight or slightly swept aft and in a steep dive they can be folded away almost parallel to the body. The feathers play an important part in providing a very lightweight surface attached to the small skeletal bones. The fine structure can be adjusted to control the flow of air through the wing or spread to form the tail and special auxiliaries such as the alula or leading edge slot and extended tip 'fingers' which control separation and vortex generation. Feathers also perform two other important roles, i.e. insulation (against loss of body heat in cold weather or against desiccation in hot weather) and camouflage. The general design of birds employing the methods just described apply over a wide range of size, e.g. from a tit weighing a mere 0.02 lb (9 g) to the Australian crane weighing 20 lb (9 kg).

But a bird's performance depends on its life style: some glide long distances and circle effortlessly in rising thermals of hot air (vultures),† others fly short distances but need to rise vertically from grassland, others float on water and some such as puffins even dive into it and swim fast beneath the surface. In all cases the size, shape and motions of the wings and feathers are specially adapted to the requirements of speed (up to 100 mph (45 m/s)), distance (up to 3000 miles (5000 km)), environment, etc. Perhaps the owl has the most interesting speciality, viz. special feathers and wing edge shape to minimise vortex generation so permitting noiseless flight so he can sneak up on prey in the stillness of the night.

So far the exterior shape of birds has been described; there are also some interesting internal features. Firstly the weight must be extremely light, e.g. the 7 ft (2 m) span frigate bird's skeleton only weighs 4 oz (0.1 kg) which is less than that of its feathers. Moreover the centre of mass must also be correct and to do this a bird's skull contains a brilliantly designed hollow braced bony structure. Even a large crow's skull weighs less than 1% of the bird's total weight. Heavy teeth are dispensed with, their duty being taken over by the gizzard, situated near the centre of mass. Birds need high metabolic rates

---

† Have collided with aircraft at 13 000 ft (4 km) altitude.

which double with a 10°C rise in temperature. Compared to man's 98.6°F (37°C), sparrows at 107°F (41.7°C) and thrushes at 113°F (45°C) seem remarkably hot blooded. A consequence of this high metabolism is a requirement for a large oxygen supply which is met by supplementing the lungs by several connected air sacs distributed throughout the body and even into hollow bone structure. These assist the lungs as superchargers and provide for cooling. A flying pigeon uses a quarter of its intake air for breathing and the rest for cooling. In contrast to a man's 5% body volume for lungs a duck's respiratory system occupies as much as 20% of the total. Of this 18% are air sacs, and only 2% are the lungs proper.

Unlike the puffin, which spends much of its time under water, the flying fish can make extended glides through the air after propelling itself by means of fins at the rear. Its front fins first open to give wing-like surfaces the lift on which first raises most of the body clear of the water surface but still allowing the rear fins to trail in the water and provide propulsion forces.

Bats fly by flapping wings but use structured membranes instead of feathers. Unlike birds however in which the vestigial fingers are fixed into a single bone structure the five 'fingers' of a bat's wing are separate and clearly seen. They permit complete wing folding after flight and also are used to camber the leading edge and deflect the trailing edge to increase lift coefficients at low speed, as in modern aircraft. Some of the 875 known species of bat have specialised in high speeds and are characterised by higher aspect ratio wings (analogous to the swift) and have streamlined 'finger' bones which in others are merely circular in cross section.

Other land animals manage to glide from high trees by several means. The flying sugar squirrel and flying possum have stretched skin surfaces between their limbs. The flying lizard has membranes on either side stretched between extensions of its ribs, and the flying frog has four webbed feet. The flying snake is perhaps the strangest of all. Once launched from a branch it glides head first with its body stretching behind it in a series of bends so that a large area is exposed to the air. The cross-section of its body is not circular but is concavely curved underneath to offer a greater area and probably develop vortices in the cross flow. Perhaps these animals may not be very elegant flyers or blessed with very high ratios of lift to drag. Nevertheless the possession of this additional freedom of movement has proved essential for survival beyond the epochs when many contemporaries, less well-endowed, became extinct.

The last class of flying animal is the insects which includes large beetles weighing over 0.02 lb (10 g), bees and wasps which vary from 0.0002 to 0.002 lb (0.1 to 1 g), down to minute creatures which turn the scales at 10^{-7} lb (0.000025 g)! It turns out that the means employed by these small flyers is as different from that of birds as they are from the supersonic aircraft.

Fascinating aerodynamic experiments have been made on the flight of the locust.[26] Weis-Fogh and Jensen harnessed living locusts to a measuring balance in a wind-tunnel and measured lift and drag forces at different wind speeds and incidence, filming the wing motion at the same time. Thorax

Fig. 45 Locust in wind-tunnel experiment

temperatures were measured as an index of muscular work done (fig. 45). Other experiments revealed that the locust detects the speed and direction of the relative wind by means of sensitive hairs on its head. These remarkable experiments gave an indication of the conditions in which locusts would fly and those in which they would stay on the ground. They were carried out with great skill and care, and were directed to a better understanding of locust flight in relation to rising air and winds, as part of an international effort to control locust plagues in Africa and the Middle East.

The essential difference between bird and small insect flight is best shown by the means they adopt for hovering. A hawk hovering in a breeze essentially employs normal wing motions but all surfaces, such as tip and tail feathers are extended and curved, and the sweep of the motion is exaggerated. The humming bird (0.011 to 0.044 lb (5–20 g)) which can hover in zero wind does so by orienting its body nearly vertically, so the wings which in normal flight move nearly vertically now move in near horizontal paths. By changing incidence at the extreme point of the motion, a lift force is created both during the forward stroke and in the backward stroke, and the wing leading edge remains the first line of contact with the air throughout the whole motion. Its feathered covered wings, flapping at a rate of between 15 and 50 cycles per second, generate a lift coefficient of 2. Some moths also hover in this way.

Weis-Fogh,[27] continuing his work on insect flight was puzzled by the ability of very small size insects to hover since the Reynolds number is extremely low. The Reynolds number for the 'normal hovering' employed by the humming bird is about 15 000; for wasps it is 4000; small fruit flies about 200 and for tiny parasitic wasps as low as 20–50. It is known from measurements and the theory of aerofoils, that lift generation falls off markedly at low Reynolds number as the boundary layer becomes relatively thick and circulation development is hindered. Even with the rapid wing beat of 400 cycles per second enough lift could not be generated to support even its minute weight, were it produced by the same means as that of the humming bird. Weis-Fogh refined experimental techniques of employing a very fast ciné camera operating at 8000 frames per second, to show that the small insects employ a totally different arrangement of wing beats called the 'clap fling' (fig. 46). With the body nearly vertical, the two wings are brought into

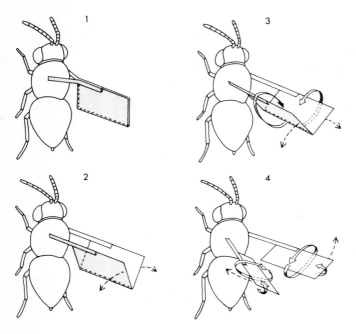

Fig. 46 'Clap and fling' flight of the wasp *Encarsia formosa*. At 1 the two pairs of wings are held firmly together as a vertical plate and in a clapped position. At 2 the wings are flung open at high speed so that the widening triangular gap between them must be filled by air which flows in (3). In 4 the right and left wing surfaces separate and begin their horizontal sweep, each side carrying the vortex of air formed during the fling. (From 'Unusual mechanisms for the generation of lift in flying animals' by Weis-Fogh. Copyright © 1975 by Scientific American, Inc. All rights reserved.)

flat contact behind the back ('clapped') and are then flung open relative to each other as one would open up a book. This action forces air from outside to rush over the opening edge to fill the vacuum. Once the air between them has reached about 90° they are then swept horizontally, carrying the vortex circulation with them that the flinging open movement created. This rapid unsteady creation of a lifting circulation is highly successful, as the lift coefficient is estimated to be as much as 5. This method of creating impulsive lift at low flight speeds is now believed to be employed by butterflies, and even by pigeons and ducks when taking off rapidly.

Yet another mechanism is needed to explain the hovering of dragonflies and hover flies, whose bodies remain essentially horizontal. Here a different 'flip' mechanism has been suggested, recognising that the leading edges of the transparent gauzy wings are rigid, and the remainder is flexible. In returning downwards after the completion of the upstroke, the whole leading edge is deflected rapidly downward, the rear part of the wing being unaffected. Much more research will be done in the future on animal flight and much remains to be learnt: there are even smaller insects suspected of having ways of flying yet to be discovered.

In summarising animal flight, it is inevitable that comparisons and contrasts will be made with aircraft flight. It will be noticed that aerodynamics cannot be discussed in isolation from other elements of the animal. The wing shape determines the structural layout, the strict demands of weight reduction leads to special bone construction and feather design. Very rapid control of flight direction is met by many swift acting joints and energy for propulsion is influenced by blood temperature, heart rate, rich diet and efficient oxygen intake and cooling. Clearly aerodynamics has an overriding influence in deciding what overall shape is best, but without well matched and equally enterprising structure, control and propulsion the exacting demands of flight will not be attained. And so it is with aircraft, as we shall see. Animal flight shows how a 'classic' solution, e.g. the bird can be adapted by subtle changes of shape and construction to operate over a very wide range of performance. In aircraft we see similarly the evolution of classics such as the Boeing B47 jet bomber and its many descendants. Means for hovering also require totally different solutions such as the helicopter and jet lift design.

Model aircraft have Reynold's numbers closely similar to those of animals and their aerodynamics is consequently different from that of full size aircraft.

Man powered flight

The aerodynamics of man himself concerns air entering nasal, throat and lung passages during breathing, sneezing and coughing, the conveyance by air of pollens and gases to the smell organs of the nose, the cooling effect of wind on the skin and the perspiration process. The air drag of a human body is important in riding of bicycles and parachuting. Wind-tunnel measurements have been made of this and indicate typical values at 100 ft/s (30 m/s) to be 110 lb (489 N) when facing the wind, reducing to 15 lb (67 N) when lying down. The ejection of pilots from aircraft at very high speed exposes them to an air blast capable of tearing face tissue and breaking bones by the shock force. In supersonic aircraft ejection methods, therefore, the pilot is encased in an airtight capsule.

The idea of a man propelling himself through the air on wings is almost as old as man himself. Whether the legend of Icarus and Daedalus had any basis in a factual flying attempt is unknown. Suffice it to say that the great Leonardo da Vinci produced sketches of a device that a man could lie in and operate bat-like wings by means of levers. Little more was heard of the concept until the 1930s, when two designs (one Italian and the other Austrian) were towed into the air. In 1959, however, the Royal Aeronautical Society in England set up a competition with a £3000 prize, put up by Henry Kremer. There followed many enterprising designs and as many failures, until the American Paul MacCready employed some clear thinking principles of design and aerodynamics to produce the Gossamer Condor which gained the prize in 1977–by which time the prize had grown to £50 000.

Although a man can exert well over one horsepower in pedalling for a few seconds at take off, the steady power thereafter must be less than $\frac{1}{2}$ hp (0.37 kW) and probably only a quarter. Light weight is essential, and low speed, but to provide enough lift for even the light weight of 210 lb (941 N) (only 70 lb (313 N) for the craft itself!) at the slow speed of 11 mph (5 m/s) needs an enormous wing, 720 ft^2 (67 m^2) spread over a span of 96 ft (29 m) as great as that of the Boeing 727 jet airliner. MacCready's secret was the large light wing with a wing loading of less than 0.3 lb/ft^2 (14.4 Pa) (by comparison the Wrights' biplane had a wing loading o 1.5 lb/ft^2 (71.8 Pa)). This big structure could only be kept light by using struts and wire tension bracing. As many as 70 bracing wires from 0.022 to 0.035 in (0.6 to 0.9 mm) diameter made this possible, and although they created an air drag it was not high at the very low airspeeds used. By housing the pilot and pedal gear in a gondola-like fin beneath the main wing, unnecessary drag was eliminated. The large lightweight slow-turning propeller behind was specially contoured for the low speeds. By extension of the same principles, a second aircraft–Gossamer Albatross–flew the English Channel (22 miles; 35 km) in 1979. After many centuries, Icarus had been vindicated–but would a new sport emerge with hundreds of Icari flying all over the countryside? Such aircraft are inevitably flimsy, they are sensitive to anything but the lightest breezes, and there seems little hope of increasing their aerodynamic efficiency much further. Nevertheless, it is exciting to think that such an achievement was made between the first and second editions of this book!

Another aerodynamic effect in human sport is the drag experienced by skiers engaging in international contests. Since as little as a hundredth of a second can separate first and second place winners, scientific tests have been made in wind tunnels to measure the air drag of different stances on skis. In tests done in Canada's National Aeronautical Laboratories' wind tunnel, at 50 mph (22 m/s) the 48.5 lbf (216 N) drag of an erect skier was reduced to a mere 13.5 lbf (60 N) by adopting a crouching 'egg' shape. Further reduction below this figure will be difficult, but various refinements to reduce protuberances of buckles and straps has been attempted.

Boomerangs

The Australian boomerang is a traditionally L-shaped (or banana-shaped) swept back wing, flat on one side and curved on the other. Thrown by hand, it can be made either to fly straight for killing animals, or in war, or to fly a curving path to return to the sender as a sport. Although also known in Egypt and India, its remarkable flight has been somewhat of a mystery and subject to much speculation and some scientific analysis, e.g. Walkers' paper to the Royal Society of London in 1897. Felix Hess reported some excellent photographs of actual flight in 1968 and also made remarkably accurate computer simulations as well.[28] Size is not critical and is typically 10–30 in (0.25–0.76 m) across the tips. Shape is, in a sense, critical since, for any planform, the camber and aerofoil shape should be appropriate, but many other designs seem to work quite well, including orthogonally crossed straight arms, and even an eight-sided 'star'. A boomerang leaves the thrower's hand at a speed of about 60 mph (27 m/s) rotating at about 10 times a second. It should initially be held nearly vertically with its flat surface away from the (right-handed) thrower and the curved surface pointing towards his left. Thus, the spin is initially about a horizontal axis. The advancing blade meets the air with its rounded leading edge (as in an aircraft wing) and generates lift. As it retreats backwards from the general speed in the line of flight its lift reduces and may even disappear. The other blade then advances in a similar way. Hence, the lift is not distributed evenly over the circular area swept by the rotating blade, but acts mostly in that half where the blades move forward in the same direction as the velocity of the boomerang itself. This unbalanced lift creates a torque which, acting on the rotation of the boomerang causes it to process as a gyroscope does. Thus, after a period of direct flight away from the thrower it tilts over towards the horizontal, begins a circling arc, and eventually lies horizontal, spinning about a vertical axis. This continues until its forward velocity is lost, and it descends nearly vertically at the feet of the thrower. That occurs if all is correctly designed and thrown. Most amateur flights are miserable failures, but the record flight stands at 108 yards (99 m) out and return!

The more popular Frisbee, a saucer-shaped plastic disc which is thrown at a small angle of incidence to its flight path, also works because of a rapid spin imparted at launching by a flick of the wrist. Its behaviour is easier to appreciate than a boomerang, which has the complication that its centre of mass is offset from the surface and creates very asymmetric lift forces. To throw a Frisbee correctly to the desired point in a given windspeed, six quantities must be correct when launched, viz: (i) angle of elevation, (ii) azimuth angle (direction as in a compass), (iii) speed, (iv) rotational speed, (v) angle of incidence to the flight path, (vi) angle of roll (bank as in an aircraft). If the wind changes, all will need adjustment. By comparison, if a tennis ball is thrown, only three quantities can be changed, e.g. (i), (ii), and (iii).

Cricket balls, on the other hand, can be made to develop significantly curving flight paths by very careful spins, and by setting the seam line at an angle to the line of flight by deft handling by the bowler. Spin can manifest itself in many ways. If the ball spins about an axis roughly parallel to its line of flight, it 'breaks' on pitching by frictional contact with the ground. If the spin axis is at right angles to the flight, a Magnus force (chapter 3) will be developed, and the lateral aerodynamic force will curve the path of the ball through the air. The deviation can be either up, down, or to either side. A more subtle effect was recently described by Doctors Mehte and Wood.[29] The air flow passes smoothly over the front of the ball in flight, but becomes turbulent and separates into an eddying wake just after it has passed the maximum diameter. Thereafter the wake has a width depending on the exact point of separation. Where this point occurs depends on the Reynolds number, i.e. the speed, and also on the position of the seam which can prematurely 'trip' the laminar boundary layer and make it turbulent. If the bowler aligns the seam at an angle to the airflow, and does not spin the ball, the separation can be made asymmetric. Because the downstream pressures are then different on either side of the ball, a swinging force is produced. If the ball is bowled too fast, i.e. above about 70 mph (31 m/s) the asymmetric boundary layer separation does not occur.

The dimples on golf balls were found accidentally to reduce air drag and increase flight distance, and this has been later proved in actual tests. It has also been discovered that minute grains of sugar beet pollen, only 4×10^{-4} in (0.01 mm) in diameter, also have a dimpled surface. In nature this is carried by the wind. Whether the aerodynamic effect is the same as that of a golf ball is not obvious as it is unlikely that the pollen is impulsively launched as is a golf ball. Moreover, the aerodynamic scale (Reynolds number) is different by a factor of about 3000.

Airships

While the aeroplane imitates the birds and the helicopter the hoverfly, the airship and its associates appear to be a quite unique invention of man. The commonly adopted name for the species of buoyant devices is LTA or Lighter-Than-Air to distinguish them from aircraft which are termed Heavier-Than-Air–although I have never come across the initials HTA in this connection. Balloons, lifted up either by hydrogen gas or hot air, have existed since 1783 and the much larger dirigible framed airships and Zeppelins were popular in the first three decades of this century. The reason for including the subject here is not for historic or nostalgic reasons but because there are stirrings of a revival of this class of aerial transport.[30] However, in spite of the emergence of many interesting and novel design proposals there are probably no more airships flying around today than there were 20 years ago. Many new uses are offered which include: relatively quiet V/STOL in and out of city centres; lifting very large and heavy objects [c. 500 tons (508 Mg)] and transporting them; patrol and surveillance of 200 mile (322 km) sea limits;

logging operations in tropical forests, and as an energy efficient form of aeronautical transport generally.

The primary source of lift of an LTA is the weight of the air it displaces. The gas within is there to equalise the internal pressure with that of the outside air. The lift is then the product of its volume and the difference in density between the gas (usually helium or hydrogen) and the air. As density difference is small and requires 14.1 ft^3 (0.88 m^3) of hydrogen or 15.2 ft^3 (0.95 m^3) of helium to lift 2.2 lb (1 kg) at sea level, airships are extremely large; the German 'Hindenburg' of 1936, for example, had a capacity of 7.1 million ft^3 (2 × 10^5 m^3) and was 803 ft (245 m) long. Airships are usually cigar shaped to minimise drag but the power required for propulsion increases as the cube of the speed which rarely exceeds 150 km/h (the Hindenburg's maximum speed was 88 mph (39 m/s). The Reynolds number for such an airship is 6.5 × 10^8 and the boundary layer thickness at the rear could be as much as 20 ft (7 m). The big streamlined hulls are unstable alone and have aft fins for directional and pitch stability. Moving surfaces at the rear (and perhaps in the future also at the front) provide aerodynamic controlling moments to steer the airship around turns, to climb or dive or to overcome the dynamic disturbing effects of wind or storm gusts. Additional aerodynamic lift is sometimes obtained by flying with a nose up angle and this method has been used in taking off along a runway at overload. New designs, Dynaships, apply spreading of the buoyant volume laterally to achieve a wing like shape and improved dynamic lift. Large aerodynamic forces are exerted on the envelope and through into the hull structure: some structural failures of early airships were attributable to neglect of these pressures. But although the foregoing aerodynamic effects are vital to the design and operation of airships it must always be remembered that they are buoyant devices and they must obey fundamental laws of aerostatics whatever else they claim to do. An unfortunate feature of aerostation is that the lift is unstable vertically. A light airship will continue to rise and a heavy one will continue to fall until either gas is valved off or ballast is jettisoned or the so-called pressure altitude is reached. Maiersperger has pointed out that a pure airship must sacrifice 1% of its gross lift for every 100 m rise in altitude and should carry a minimum of 3% gross lift as ballast. Furthermore, lifting gas is frequently diffused by 5% of air hence the total penalty incurred by flying at, say, 1640 ft (500 m) is 13% which represents a serious defect of payload capacity. Temperature effects are also significant. If the full gas cells were heated up by 40°F (22.2°C) passing from clouds into clear sky can induce such changes, requiring height or ballast adjustment. Now some of these characteristics may not be inconvenient, e.g. in sea patrolling where altitude may not be critical, or in carrying heavy loads short distances at selected times. Instances where such effects would be less tolerable would be on transcontinental flights over high mountain ranges 9840 ft (3000 m) or maintaining timed schedules in all weathers. Airships have been designed to fly above 20 000 ft (7000 m) but they have to be much larger and more costly. A typical volume increase required to operate at 16 400 ft (5000 m) instead of at sea level would be 64%.

The efforts directed towards the revival of the airship can be traced to the late 1960s. Essentially the stimulus has come from the existence of today's more advanced technologies of lightweight and composite materials, refined methods of structural design, sophisticated control systems, microelectronics, better weather forecasting, radar and lightweight propulsion systems–to name but a few. Great originality has been displayed in designing alternative shapes of airship, such as multihulls, lenticular saucers and delta winged creations; gas turbine engines driving helicopter blades, whose axes can be turned to provide either horizontal or vertical propelling forces, and in more responsive and clever controls. Some of the more intriguing concepts will be described but the essential aerostatic characteristics, the limited uplift provided by helium gas and the inevitable aerodynamic consequences still require solution.

Better control of buoyancy lift is clearly a primary aim especially near the ground where turbulence is high and where too high a rate of descent would lead to damage. Brennan has proposed using the powerful civil jet engines to deliver large quantities of compressed air into ballonets as ballast. As much as 20 tons could be supplied in a minute. Several ballonets arranged along the hull would also permit longitudinal trim control. Another advantage inherent in having a large compressed air supply is to inflate landing bags beneath the airship to cushion the landing. The hybrid design incorporating large helicopter rotors offers great improvement in height control since the aerodynamic forces of the rotors can be virtually independent of altitude or speed. Furthermore, thrust forces can be applied almost instantaneously with modern control systems unlike the classic airships which typically required 30 seconds to apply full elevator deflection. The manoeuvring forces available in the helicopter-airship hybrid would be an order of magnitude greater than these available in the 1930s and might well dispose of most, if not all, of the problems of lack of control experienced in loading, near ground operations and the effect of solar heating on buoyancy. A typical project is the Piasecki Helistat which combines a 5.7 million ft^3 (1.61×10^5 m^3) airship with four CH-53E helicopters. But the mixing of the two systems may give the hybrid excessive deadweight so the payload fraction could be uneconomic. Although there are many other enterprising and original ways of combining gas volume with rotors none has so far yet flown to prove whether such systems herald a new era of aeronautics. Another hybrid which seems to violate engineering sense is one combining the large streamlined airship hull with large wings of a conventional aircraft. The Megalifter concept of 1974 employed the wing and engines of the vast Lockheed Galaxy transport aircraft with an airship body and twin finned aircraft-like tail. A 200 ton (203 Mg) payload was estimated. A major difficulty is bringing together the large concentrated aerodynamic lift of the wing to join with the spindly distributed structure around the hull. Uneconomic weight increase would seem inevitable. Other hull shapes designed to capture significantly more aerodynamic lift than the conventional hull are shaped as large delta wings somewhat like the space shuttle. Although in steady flight the aerodynamic and aerostatic lifts could mutually benefit each other, structure weight and precise control near the ground would require careful and lengthy investigation. Some less radical improvements over conventional airship performance have been offered by schemes to apply boundary layer control to reduce drag by sucking in the thick boundary layer near the rear of the hull and expelling the air through the power unit. Various nuclear propulsion ideas have been published but have not been received with much enthusiasm.

There remains a great deal to be done to make any of these quite new concepts come to fruition. Although for a large airship the fuel consumption per unit payload-distance may be about one-tenth of that of an aircraft the much lower speed results in the economics of productivity of the LTA being inferior in most cases. It cannot therefore be expected that new classes of airship will gradually replace present day airliners. Some specialised tasks, some of which have already been mentioned, such as 500 ton (508 Mg) loads, could well be served by airships and to succeed in development will require the solution of several aerodynamic problems.

Aerodynamic problems of the aeroplane

In an aeroplane there are seven basic features which are determined largely by aerodynamic considerations.

(i) *Provision of enough lift.*[31] The lift is produced by the wings, downward-deflected jets or rotor blades. The lift on a fixed-wing aircraft is the result of its speed as well as a circulation flow induced by viscosity and an asymmetry of shape. The lift equals the weight in steady level flight, but much more must be available for climbing, manoeuvring and landing.

(ii) *Efficiency of lift in relation to drag.* Since lift is generated by air flowing over wing surfaces it cannot exist without drag. The correct flight speed is attained only when the propulsion engine thrust at least equals the drag. Wing design, therefore, involves a knowledge of both lift and drag and how they change with shape, incidence and moving flaps and slots. The lift–drag ratio or L/D is an important index of flight efficiency.

(iii) *Efficiency of propulsion, propellers or jets.* The propulsion system provides the forward force needed to overcome drag and inertia, giving the desired flight speeds. The first aeroplanes had propellers (or aeronautical fans) but nowadays design is more complicated as jets, turboprops, turbofans and helicopter rotors are also in use. When jets are tilted to provide part-lift and part-thrust, the aerodynamics becomes even more challenging.

(iv) *Stability.* Stable flight is possible if two conditions are satisfied: firstly, that the forces and moments should all balance; and secondly, that if the aeroplane is disturbed from a position of equilibrium (e.g. owing to a wind gust or shift of cargo), the changed aerodynamic forces and moments should automatically act so as to return the aeroplane to its previous steady state. Aerodynamics enters into this problem in many ways: for instance, the tail surfaces provide most of the stability and the forces on the tail depend on how the airflow is deflected passing over wings and bodies.

(v) *Control*. An aeroplane frequently changes speed and direction, and aerodynamic control is exercised by the pilot to place the aeroplane at the correct angle to the oncoming airflow in order to create the right lift, drag and side forces, which will then move the aeroplane in the desired flight path. Control is arranged usually by movable surfaces which change the airflow and hence the forces and moments. They must be light to operate, and must not vibrate or become unstable.

(vi) *Air-conditioning*. To keep the aeroplane efficient leaks are scrupulously avoided. 'Aeronautical draughts' are expensive faults in terms of increased drag. Special inlets are provided to collect air, and this is then heated or cooled, pressurised, humidified and finally passed to the occupants. Conditioned air also goes to the electronic equipment.

(vii) *Determination of local air pressures and heating over the surface for stressing purposes*. The overall shape determines lift, drag and stability, but the structure must be strong enough to withstand the air pressure on the surface. It must not deflect and thus disturb the smooth airflow, and it must be able to expand with heating, and without failing. For these and other structural problems, the distribution of air pressure and the heat transfer over the surface must be calculated or measured.

The shape and speed of an aeroplane determine how the air is deflected by its passage. The changes in air velocity lead to the aerodynamic pressures, forces and moments which contribute to the seven features previously described. Although some of the basic aerodynamics underlying all these features is the same, very different approximations are made in each subject and the amount of detail required for each is very considerable.

The creative task in *designing* aeronautical vehicles is to combine the many aerodynamic qualities into a whole, and this is frequently a surprisingly good compromise satisfying many conflicting requirements. The aeroplane has to perform many diverse duties. There are wide variations dictated, for instance, by the size and weight of cargo to be accommodated, range, speed, military qualities, cost and smartness, and this accounts for the many different shapes and sizes of aeroplanes. Trends or fashions in these shapes can be detected. For instance, they change gradually as speeds increase (fig. 47) (the shape of a low-speed sailplane is quite unsuitable for a supersonic fighter or missile). In a different sense, 'fashions' in the shape of aeroplanes having roughly similar performance characteristics undergo changes over long periods of time. This is a consequence of the general progress of aeronautics brought about by more efficient structures, better manufacturing methods and improved automatic devices. The aerodynamicist has not only to deal with the basic problems of lift, drag and propulsion, but also to keep in step with changes of shape and speed and with the gradual evolution of techniques.

In aeronautics, aerodynamic science appears in its most highly developed form. The reasons for this are:

(i) the need for a very efficient vehicle which cannot afford any surplus weight,

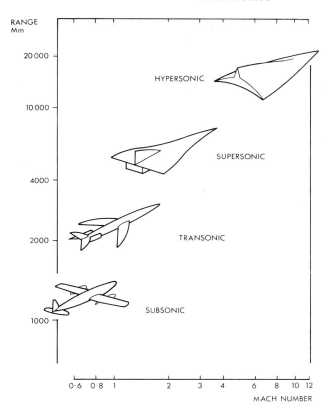

Fig. 47 The spectrum of aircraft (after Küchemann[32])

(ii) the need to pre-calculate every aspect of required flight performance and to ensure that the design will meet it,
(iii) the paramount criterion of safety, to minimise the chance of catastrophic error occurring for the first time in actual flight.

Safety is based on three essential factors: the structure must not fail, the machinery must be reliable, and the handling characteristics of the aeroplane must be correct and manageable, i.e. so that it will not get out of control in flight or be liable to landing or take-off accidents. In ground-based engineering it is often quite adequate to estimate structural loads simply, and make the structures several times (say 6) stronger than this, allowing a large margin of safety. The extra weight involved by this method is aeronautically quite unacceptable. Aeronautics can only tolerate low safety factors, usually about 1.5, and because of this many more details, of aerodynamic origin, have to be worked out in the design stage.

Both in military and civil aircraft, there is fierce competition either between companies or nations, and this has led to high priority, highly qualified staff and expensive facilities. It has been estimated that over £100 000 million were spent on aeronautics in its first fifty years.

The aerodynamics of aircraft is most conveniently described in relation to the distinct speed regimes of chapter 4, i.e. subsonic, supersonic and hypersonic. The aerodynamics of propulsion form a separate section.

Subsonic aeroplanes

The lift (L) on a moving wing can be calculated from the circulation (K) of air round it (fig. 5), viz:

$$L = \rho V K$$

The circulation strength depends on the shape of the aerofoil and its incidence and is calculable by Joukowski's condition (fig. 1) that the circulation is just sufficient to ensure that the two airstreams above and below the wing surface reach the trailing edge, and leave the aerofoil at the same speed. Theoretically it can be shown that for a wing of infinite span the lift is:

$$L = \tfrac{1}{2}\rho V^2 S C_L = \tfrac{1}{2}\rho V^2 S 2\pi\alpha$$

where S is the wing area and α is the incidence in radians. But aeroplane wings are not of infinite span and 'spill' their circulation, mostly at the wing tips, to trail downstream in vortices as shown in fig. 48. These vortices induce up, down and lateral flows round the wing which reduce its lift at a given angle of incidence to:

$$L = \tfrac{1}{2}\rho V^2 S 2\pi\alpha \frac{A}{2 + A}$$

where A is the aspect ratio = (span)^2/area. This shows that the larger A is, the more efficient is the wing, i.e. the tip vortices are kept farther apart. This is to be seen in high performance gliders in which A exceeds 20. The drag of aeroplanes (and geese) is reduced by flying in formation, lift being gained from the upwash of the others' vortices. The drag reduction can be nearly 20% for a formation of five aeroplanes.

Although the previous equation suggests that C_L increases with incidence there is a practical limit $C_{L\max}$ beyond which the air will not flow smoothly over the upper wing surface. At this point the flow breaks away and a disturbed wake of eddies appears. The wing is said to stall: lift decreases markedly and control may be lost too. Aerofoil sections are designed not only to give good L/D for high speed and high maximum C_L for slow speeds, but to give mild stalling characteristics.

Since lift equals weight in steady flight, i.e.

$$W = L = \tfrac{1}{2}\rho S C_L V^2,$$

and C_L increases with α up to $C_{L\max}$, slow landing requires a high $C_{L\max}$ and

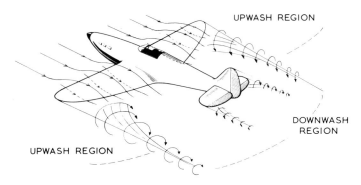

Fig. 48 Circulation and vortex flow of aeroplane wing

high incidence. Movable flaps at the rear deflect to increase circulation and slots at the leading edge speed up air over the upper surface to postpone the stall to a higher incidence.

The choice of wing section depends on many other factors as well, e.g. at low Reynolds number the thickness can be up to 20% of the chord, with the maximum thickness a quarter back from the leading edge. At high subsonic speeds when the local velocity above the surface tends to reach sonic speed it is customary to have the maximum thickness much further aft.

The flow pattern and local Mach number around an aerofoil change dramatically as speed increases through the speed of sound (fig. 49).

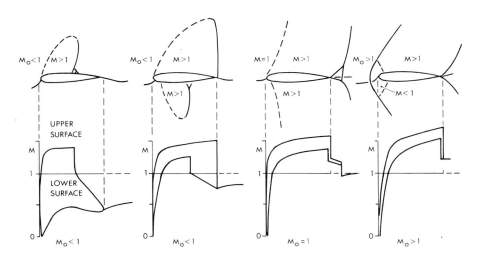

Fig. 49 Flow patterns of two-dimensional aerofoils at different mainstream Mach numbers

Swept back and swept forward wings

As aircraft speed increases further towards that of sound (transonic) the air becomes expanded as the airspeed is increased locally by the combined thickness of parts of the aircraft and can reach and exceed the speed of sound. Then, to compress back to atmospheric pressure behind the aircraft shock waves usually occur. The associated pressure changes can be large, and may shift in position rapidly, or in an oscillating manner, and can lead to dangerous changes and loss of stability and control. These effects were first encountered on the propellers of fighters in World War II, especially when diving from high altitude and were more likely to be encountered on the aerofoils of the faster jet propelled aircraft then being designed. A solution to this emerged from Germany where theoretical and experimental research conducted by Dietrich Küchemann and colleagues resulted in the swept back wing.

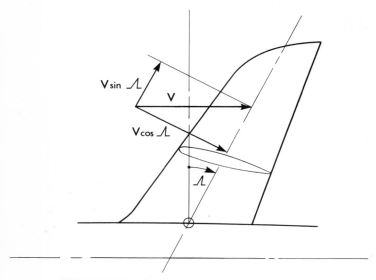

Fig. 50 Swept back wing

The swept wing can be regarded as a given structure pivoted or 'yawed' (as though about a pivot) through a sweep angle Λ (fig. 50). Then it can be shown that only the component of the airflow perpendicular to the wing centreline leads to any changes of local velocities. In other words, the pressures and velocities over the wing are those that would be found for a straight wing at a reduced aircraft speed of $V \cos \Lambda$.

Küchemann's very comprehensive understanding of the complete airflow around an aircraft (reference 32) was also applied in shaping the body in relation to the wing so that the interferences were beneficial rather than unfavourable, curving the wing tip shape, shaping the intake, and providing streamlined 'bullet' fairings between tail plane and fin to avoid transonic buffet disturbances. The effect of all these alterations in the shape of the whole aircraft (configuration is a modern term for this) which resulted from this treatment culminated in a new class of flying machine radically different in virtually all respects from its subsonic predecessor. Küchemann's fundamental method was firstly to understand the behaviour of the airflow at the appropriate speed and then seek a shape that was in all ways compatible with the aerodynamic effects. Whereas the 'classic' unswept wing shape had dominated aircraft design from Sir George Cayley's glider of 1799, through the Wright Brother's biplanes to the Spitfire/Fw 190 of World War II, the new transonic shape emerged during the 1940s and 1950s–well typified by the beautiful lines of the Hawker (BAC) Hunter (fig. 51).

Fig. 51 The British Aerospace (Hawker) Hunter

These principles apply equally whether the wing is swept back or forward, so it is natural to ask 'Why are swept forward wings so uncommon?' The answer to that lies in the effects of wing bending. If a sweptback wing bends in the usual way, the tendency is for the top to curl upwards. The effect of this bending is to reduce the airload on the top and consequently lessen the tendency for the top to curl up. It can be seen that this same type of bending would lead to *increasing* tip-curl on a swept-forward wing, which could be unstable and lead to structural failure if the wing stiffness is not increased enough for high speeds.

Colonel H. Krone, of the USA has since pointed out that there is a way of avoiding this wing tip divergence. If the wing bends in such a way that the incidence of the wing tip is not increased, then there is no instability. This can be arranged by making the wing structure in fibre reinforced plastic (such as carbon fibre) in which most of the fibres are about 15 degrees more swept forward than the wing centreline. If this is done, the designer is free to choose swept forward wings, which have some potential advantages over swept back–such as a better (more elliptical) spanwise distribution of lift, and potentially less supersonic drag for given angle of centreline sweep.

Aeroplane boundary layers are, in general, partly laminar and partly turbulent. Great efforts are now made by structural and production engineers to give large smooth unjointed surfaces to wings and bodies to reduce roughness drag, and this also encourages the transition to move aft. The concept of the wholly laminar boundary layer aeroplane suggests very attractive improvements in performance, e.g. a skin friction drag reduction of 40%, a 60% range increase for the same weight or cost reductions of up to 30%. To do this, however, requires more than a smooth surface: the boundary layer must be continually sucked away through holes or porous wing surfaces by a suction engine. This process is difficult to engineer in practice, for it has not been possible, so far, to evolve a porous surface system which is not vulnerable to blocking by insects during take off. Experimental aircraft flew with such wings in the 1950s but interest then lapsed until the energy crisis of 1973 which made the concept attractive once again. NASA has flown experimental aircraft with Laminar Flow Control in the 1970s as part of its Energy Efficient Aircraft Program. Although the completely laminar boundary layer aeroplane has yet to appear, boundary layer *control* is already employed in some military aircraft to improve low-speed landing qualities. The boundary layer thickness can be controlled either by sucking or blowing air through slits to change the airflow over flaps or slots.

In the early days of flying, many aeroplanes broke up in flight because of flutter, which is an instability caused by the interaction of air loads with the elasticity and inertia of the structure. This can now take many forms and the general subject is re-defined as 'aerothermoelasticity' to include heating effects also. One of the earliest examples was wing-torsion flutter. In this case, a wing twists from the action of the aerodynamic pressures over it, and this twist increases the lift further. When the elastic restoring action of the structure overcomes the air loads the twisting reverses sign and starts an oscillation. Below a certain airspeed, any disturbing effects are damped out, but at a 'critical speed' the damping changes sign, and above this speed the structure may disintegrate before the pilot can reduce speed. Flutter in its many modern forms is a continuing problem for the aeroplane designer, but methods of theoretical pressure calculations and laboratory testing of structures subjected to calculated loads are his safeguards.

Inflatable aeroplanes now exist which can be packed away into two small cylindrical containers. Considerable ingenuity is required to transmit propeller thrust forces into the 'structure' and maintain rigidity for the aerodynamic control surfaces. Such radical departures from previous experience bring with them several unfamiliar problems.

The foregoing are only a few of the major aerodynamic features of an aeroplane, but there are many others of quite a different kind, as the following three examples will show.

Aeroplanes and helicopters may encounter snow and ice when flying in cloud. This builds up on surfaces and in engine intakes, changing the shape and hence the aerodynamic pressures and forces. Severe icing has led to several crashes. Some features of icing can be simulated in wind-tunnels, but special test aircraft are also used to spray water droplets on to wings, engines and propellers. A typical build-up of ice is shown in fig. 52. From the results

Fig. 52 Ice formation on aircraft engine

of such tests, engines, heating systems and control surface shapes are improved. At high altitude, the water vapour in the engine exhaust condenses and forms 'contrails' in the vortex wake behind the aeroplane. Later, they disperse or are extended by winds. Contrails are an embarrassment to military aircraft, whose positions they betray, but research, conducted with blow torches in a small wind-tunnel, has evolved a method of eliminating them on jet bombers.

There is a special class of aerodynamics associated with an aircraft's armament. The gun blast ejected at high speeds can snuff out a jet engine or fracture panels and dropping ballistic bombs can be troublesome at high speeds. Bombs have been observed to 'fly' at incidence after leaving an aeroplane and hence fall away less swiftly than intended. A guided missile fired from an aeroplane launcher pylon also encounters aerodynamic disturbances as it rides through the wing or body flow.

It will be realised that aeroplane aerodynamics is particularly complex. In spite of all our knowledge and research, we can never quite keep up with the many unknowns found in new designs, and the penalties for error are severe. There have been instances of airliners having a higher drag than anticipated, with a consequent reduction in cruising speed, and the necessary changes in shape to prevent this had to be built into the airliners, after they were in service, at the constructor's expense. Such costs can mount to scores of millions of pounds.

Aerodynamics of propulsion by propellers, fans, rockets, turbojets and ramjets

Vehicles which propel themselves through the air do so by creating a propulsion force, or thrust, which is a reaction force arising from increasing the backward momentum of mass. In a rocket motor, the rearward ejected mass comes from the propellant chemicals carried within the vehicle. In the propeller, the backward momentum of the airflow is increased by addition of mechanical energy to the slipstream, and in the jet engine by addition of heat energy to a controlled flow of air passing through part of the vehicle.[33] The thrust = change of momentum:

$$F = \rho V A \Delta V$$

where A is the cross-sectional area of that part of the airstream accelerated in the reaction process, and ΔV is the overall increase in velocity of this air taken from well ahead of the vehicle to well down its wake, as shown in fig. 53.

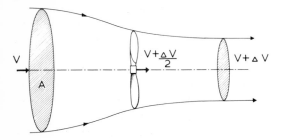

Fig. 53 The increase of velocity in the flow through a propeller

The efficiency of a thrust-producing jet of air is the propulsive work divided by the kinetic energy which is

$$\eta_F = \frac{1}{1 + \frac{\Delta V}{2V}}, \quad \text{the Froude efficiency.}$$

When ΔV is zero this is a maximum, i.e. when the jet exhaust relative to the aeroplane is equal and opposite to the forward flight speed and the wake has no kinetic energy in it. (This simplified approach does not take into account turbulence in the jet.)

Types of propulsion system have evolved as aeroplane speeds have increased. The *propeller* is essentially a group of twisted aerofoils, each section of the blades of which travels along a helical path, generating lift and drag forces. Circulation and vortex calculations give the components of thrust and torque for all parts of the blade. The usefulness of the propeller is limited by the airspeed at the tip reaching the speed of sound causing the aerodynamic lift-drag efficiency to fall seriously. Typical propeller numbers might be: flight speed 450 mph (201 m/s), diameter 12 ft (3.7 m), true tip speed 580 mph (260 m/s) ($M = 0.77$), hp input 5570 (4150 kN), thrust 4000 lb (17.9 kN), efficiency 86%, slipstream velocity increase at propeller 16%, and downstream 32%. The maximum level flight speed of a piston-propeller aircraft was 480 mph (215 m/s) ($M = 0.64$). All-supersonic propellers rotating at much higher speeds have been tested, but because of the much lower lift-drag ratio of the blade sections they have not proved economic. If a propeller is driven by an internal combustion piston engine, its waste heat must be dissipated to the air either directly or via a water radiator. Although propellers have yielded pride of place to the jet, they are now used extensively on helicopters, hovercraft and small aircraft. Small models of complete aircraft are equipped with propellers to measure overall lift and drag in the presence of slipstream, or on flying-boat models to show also the interference between slipstream and water spray. Full-sized propellers and rotors are often tested in very large tunnels, driven either by aircraft engines or electric motors. The basic theory of the propeller applies also to the turbine, fan, helicopter rotor and windmill with, of course, appropriate changes.

In the jet engine, heat energy of fuel is released in an airstream which has been slowed down and whose pressure has been increased either by an expanding duct (ramjet) or by an expanding duct and a compressor (turbojet). A typical jet consumes 100 tons of air/min, of which only about a hundredth is burnt with the fuel; the remainder helps to cool the engine and increase the mass flow that is accelerated. Jet engine aerodynamics includes the *design of the intakes*, which may be separate circular pods well away from the wing or fuselage on struts, or carefully buried within the wing or body; *the flow into the compressor* which consists of several many-bladed fans; *fuel combustion* and *heat transfer* processes in the combusion chamber; and *flow through turbine blades and in the exhaust duct*. The aerodynamics involves many thermal, density, pressure and Mach number changes as shown in fig. 54, but the aerodynamics has proved to be easily calculable mainly because the airflow is made to flow within pipe-like passages. Modelling laws are applicable to the airflow and also combustion processes. In addition, the different components can be tested separately in special rigs (fig. 2). Heat transfer to the turbine blades is a critical consideration, for with the high centrifugal stress, the metal is working very near to its strength limit. Some

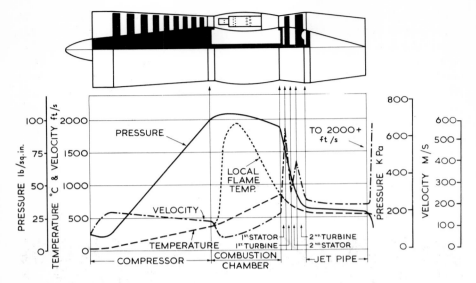

Fig. 54 Changes in the airflow through a typical jet engine

turbine blades have internal cooling air passages.

The effect of the jet engine intake and exhaust flow on its aircraft can be measured in a wind-tunnel, and model techniques with cold and hot jets have been evolved.

The added velocity imparted by the jet (ΔV) is higher than in the slipstream of a propeller of the same thrust, as the diameter of the jet is so much smaller. To improve the jet's efficiency, i.e. by lowering the average ΔV, the jet either drives a propeller (turboprop) or a ducted fan in which the cross-sectional area of the jet exhaust is larger. In the fan-jet, some air by-passes the combustion chamber into an annular duct surrounding the rest of the engine. Fan blades mounted within this duct are connected directly to the turbine, and at the exit the interior hot, fast, jet exhaust merges into the outer, slower and cooler, ducted fan slipstream. The average kinetic energy loss in the combined jet is less than in the pure jet, and the noise is also less because the shearing velocity at the edge of the jet is reduced by stages (fig. 55).

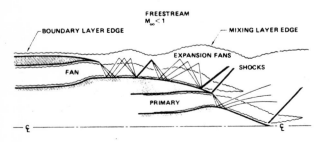

Fig. 55 Typical flowfield of a multistream nozzle

Turbojets work well up to Mach numbers of about 3, and, although experimental models have been made for higher speeds, none has so far flown. The compressor's task is to raise the inlet air pressure to a high value for efficient combustion. The inlet duct does this as well and, as the Mach number increases, the inlet pressure rises. Above a Mach number of about 1.5 the compressor (and the turbine) can be dispensed with, and an engine designed on this basis is called a ramjet. These are very simple engines with no moving parts, which can take vehicles from low supersonic speeds up to Mach 5 at least. They do not work efficiently subsonically, however, and must be combined with rocket or turbojet to cover these speeds.

The rocket, by being independent of an airflow, provides roughly the same thrust at any height and speed. The thrust improves with height as the air pressure hindering the exhaust lessens. There are aerodynamic problems associated with the interaction of the rocket jet and the wake flow (fig. 56)

Fig. 56   Missile rocket exhaust flare at altitude

and the need to minimise contact between the hot exhaust gas and the rear metal parts of the vehicle.

It is worth noting that, whilst in aeronautics the thrust is of major interest, in heating and cooling fans, and in blow torches, interest lies rather in the velocity, mass flow and temperature of the jet itself.

Rotorcraft[34]

A fixed wing aeroplane has to fly fast to stay up. In spite of the extra lift provided by flaps, slots and boundary layer control landing speeds are as high as 50–150 mph (22–67 m/s) and concrete runways have been getting larger and more expensive and swallow up more land every year. The advantage of VTOL (Vertical Take-Off and Landing) aircraft is that they can operate from very small areas, not much larger than a tennis court, in cities and on tops of buildings. The idea is not new, for the helicopter (sketched by Leonardo da Vinci in the fifteenth century and first flown by a Frenchman in 1907) meets these requirements: because lift is obtained by rotating long aerofoils about a central vertical axis, it is not dependent on the forward speed of the vehicle.

The rotor or rotors are power driven from piston or gas turbine engine and the angle of incidence of the blades is controlled in two ways. To roughly equalise the lift created over the circular area of the rotor the incidence of the blade is changed as it performs one revolution, or cycle. As the blade advances in the direction of motion of the helicopter the total air velocity over the blade increases and its pitch (incidence) must be reduced. 180° of rotation later, as the blade is retreating in the direction opposite to the flight speed, the airspeed over it is lessened so that incidence must now be increased to compensate. This arrangement is called cyclic pitch and is provided by mechanical means. The blade angle is an average value when the blade is either pointing directly forward or directly aft from the hub. Two different aerodynamic problems are encountered by the blades in each rotation. At the high speeds with Mach numbers approaching unity the blade efficiency is reduced by compressibility drag, and at the minimum speed the aerofoil may become stalled because of the high incidence. It is the conflict between advancing and retreating blade aerodynamics that limits the speed of the pure helicopter to about 200 knots (103 m/s). The other blade control is called collective pitch, which increases the pitch angle of all blades equally (in addition to whatever the cyclic pitch requires), thereby increasing or decreasing the rotor lift as a whole. So, when taking off vertically at virtually zero forward speed, the collective pitch increases and the cyclic pitch control is negligible. At very high forward speeds the cyclic pitch angle control would dominate. The fact that the lifting surfaces are rotating and have to be controlled makes the helicopter more expensive to purchase and maintain than a fixed wing aircraft of similar weight. The rotating machinery has a high air resistance, often 25–30% of the total. The whole scope of the helicopter has been transformed in recent years by ingenious new means of attaching the blades to the hub with simpler ways of controlling the cyclic and collective pitching motions. The number of parts in the hub mechanism has been reduced from typically 400 to less than one hundred with consequent reductions of maintenance actions and costs and noticeably reduced air drag. A rotor of a medium sized helicopter is about 40 ft (12 m) in diameter; it rotates at six revolutions per second, the tip speed is limited to 0.7 of the speed of sound and the centrifugal force at the tip is 400 g. Helicopters have to manoeuvre rapidly in pitching and rolling to 40° and 80° respectively and up to pitching and rolling rates of 40° and 50° per second.

The helicopter is an increasingly useful aircraft in both civil and military roles. It transports both people and goods, delivers rockets or bombs against targets in the land battle, is a valuable aid in anti-submarine warfare, has a unique rescue role in peace and war on land and sea, is used to spray crops and lift large or awkward objects into inaccessible places. The technical problems are to keep weight low, to counter the torque created by the power driving the rotor and to control the flight precisely and safely. Torque is either balanced by a vertical rotor at the tail whose controllable blades can be altered to provide lateral force in either direction, or by mounting two main contra-rotating rotors in such a way that their opposing torques cancel out. They may be mounted one above the other or at either end of the helicopter. Vibration and noise are problems which are receiving a lot of attention and at the present time helicopter aerodynamics is a very active subject. Helicopters have typically twice the drag of an equivalent aircraft and the desire to fly ever faster than before has brought about a great deal of enterprising research, new mechanisms, blade construction and shape and very advanced designs of helicopter. Some of the aerodynamic issues will be illustrated by reference to recent research findings.

The objective of advanced rotor systems is to employ new, strong lightweight materials in combination with novel manufacturing techniques to improve the lifting efficiency of the rotor both at hovering and in high speed flight. The new blades can be more tapered, incorporating advanced aerofoil sections with camber and twist, variable leading edge camber, trailing edge flaps and even variable diameter rotors. The conventional rotor with cyclic and collective pitch variations is extremely inefficient with an overall lift/drag of three or less. A theoretical maximum of 15–16 might be realised if each part of the blade could be set at its best angle. If the advancing blade half of the rotor could develop lift without the restraint of balancing the smaller value on the other side the lift/drag could be doubled (about six) even with practical blades. In the Advancing Blade Concept (ABC) unbalanced rotors are mounted one above the other on the same shaft. Each then achieves its best lift on the high speed side leaving the other rotor to counterbalance the overturning moment. A quite different principle is employed in the jet flap rotor. The blades are hollow and compressed air is supplied from the hub, travelling towards the rotor tip and exhausting to atmosphere through a nozzle all along the trailing edge. The high speed jet flow increases circulation over the rotor section which is then largely independent of the angle of incidence, thereby avoiding stalling the retreating blade. Experimental measurements confirm calculated improvement in lift of between 80 and 300% at a given airspeed, thereby permitting a significant increase of forward speed. A variant of this idea is the Circulation Controlled Rotor in which the rotor blade is elliptical in cross-section. This principle leads on to the dual blowing arrangement in which blown slots are provided at leading and trailing edges. The cyclic control is now performed by jet blowing the trailing edge for most of a blade revolution. However a second slot in what is normally the leading edge is blown on the retreating blade; here the slot is now in the trailing edge, as far as the air is concerned. Because of the high lift developed by this method rotor tip speed may be lowered, so reducing Mach number tip losses and, it is alleged, eventually giving flight speeds up to 400 knots (206 m/s). By stopping a four-bladed rotor so that it looks like an X when viewed from below the blades can then act as aircraft wings. In all the blown schemes the hub is driven by mechanical power: in earlier projects of the

1950s compressed air was used in hollow blades to power pressure jets at the tips to actually propel the rotor. This neatly avoided torque reaction but the high noise level was unacceptable.

All aeronautical vehicles have to resolve difficult design compromises between low and high speed flight and the helicopter is no exception. Various hybrid systems have been or are being developed to increase flight speed. The compound helicopter also has a main aircraft-like fixed wing, which although adds weight, which penalises take-off, offloads the rotor at high speeds. It may also have additional jet propulsion or propeller independent of the rotor drive. The tilt wing takes this idea further but at the expense of extreme complication. Other proposals are to stop and fold the rotor at high speed; or stop the rotor (of three blades) so that one lies forward along the body and the other two act as highly swept thin aircraft wings. A more recent and less complex arrangement than the foregoing hybrids is the tilt rotor brought to successful flight test in 1977 by Bell Helicopter Textron Co. of Texas, USA. The XV-15 is essentially a high wing aircraft with conventional tail. At each wing tip is mounted a large turboshaft engine contained in a nacelle driving a large three-bladed rotor. The engine-rotor system is pivoted at the wing tip and, with the rotor horizontal, it takes off vertically as does a helicopter. Once clear of the ground the rotor-engines tilt forward progressively until they act as large propellers and the XV-15 then behaves as it if were an aircraft. By this means high speeds are attained, propulsive efficiencies are high and noise is low. The control system is required to perform both aircraft and helicopter functions. If this experimental aircraft proves successful a whole range of larger transports could evolve greatly increasing the scope of VTOL at high speed and in city environments.

Reduction of helicopter noise could considerably enhance their use not only near communities but also in military flying. Internal noise which deters passengers and has been occasionally intolerable for crews is also relevant. The noise is created from several sources, viz. the rotational (aerodynamically generated) noise from main and tail rotors, wake vortex noise, jet exhaust and gearbox drives to rotor and tail. Treatment includes increasing the number and chord of blades (to reduce rotational speed), improved tip shape (which generates quieter vortices), correct separation of main rotor boom and tail rotor and detailed attention to machinery and combustion noise. Insulation and mufflers are also used as secondary cures. However the design changes to reduce noise cost money; typically a 20 dB reduction of noise could increase direct operation costs by 25% if it were applied retrospectively but if included in the original design the penalty is nearer 5%.

Aerodynamic testing on models in wind tunnels and in free flight has been developed to a remarkable degree of sophistication. Boeing, for example, test rotors in conventional wind tunnels and have developed a universal helicopter model for experiments in a special V/STOL tunnel. At Cornell a moving truck is used in free air; at Princeton there is a moving test track in which a model is moved horizontally and vertically to subject it to correct airflow direction and speed. Great ingenuity is employed in constructing scale size blades in carbon fibre to have correct dynamic values. NASA employ a control line technique with a rotating crane. Radio controlled free flight models are often used.

The increasing employment of simple and compound helicopters reflects not only their great utility but greatly improved performance, cost and acceptability. Much of this advance is of aerodynamic origin and many further improvements must be expected in the future.

V/STOL (Vertical/Short Take Off and Landing) aircraft (non-rotor types)

It had already been observed during the early development of the jet engine in the 1930s that because the engine thrust was large compared to its weight it might be possible to lift the aircraft off the ground by means of the engine thrust alone, i.e. without wing aerodynamic lift and therefore from a standing start. Active steps towards this new class of jet-lift aircraft, capable of VTOL flight, had to await the improved jet engines of the 1950s which had sufficient margin of thrust over weight to more than compensate for various aerodynamic interference losses–the margin is typically 15–25%. This is simple enough to state as a principle, and there are very many possible means of achieving this in practice, but most attempts have suffered from various drawbacks, usually overcomplexity, so that in fact only two small military VTOL aircraft have gone into actual operations, viz. the British Aerospace Harrier[35] and the Soviet Union's YAK 36. No civil type has flown beyond the experimental stage.

The first experimental aircraft to demonstrate the principle was the Rolls-Royce 'Flying Bedstead' which mounted two 'Nene' jet turbines in a framework so that their jets exhausted downwards. The pilot sat on top of the contraption–in the open air. The 'aircraft' was so elementary that providing enough jet lift to overcome the weight was an easy matter; controlling it in a stable manner whilst hovering was more difficult. In the hover this jet lift aircraft was neutrally stable, having no preferred attitude so controls had to be provided. Compressed air from the engine's compressors passed along horizontal pipes that ended in downward pointing jet nozzles. The flow of air to these nozzles was controlled by the pilot to keep the Bedstead horizontal. Soon after this successful proof of the principle many experimental designs appeared. For completeness the basic categories will be described briefly since they raise aerodynamic problems of several kinds. Rotors, rotors with wings and tilting rotors have already been described under rotorcraft in the previous section; those described here all employ jet propulsion of some kind; application to military types is assumed.

If only one type of engine is employed, which is used in all modes of flight, there are three ways to provide jet lift. One method is to mount the aircraft vertically on some form of launching carriage and let it fly straight upwards; this is called the tail sitter or VATOL (Vertical Attitude TOL). The aircraft is simplified to some extent but landing and take off are inflexible since special landing apparatus is necessary and operational landings could be vulnerable. Pilots do not like to be on their backs while slowly lowering an aircraft on to a hook! This objection is overcome by the 'nutcracker' in which the fuselage is hinged in the middle so that while most of the aircraft, engine and exhaust are vertical the cockpit is horizontal and the pilot sits in a normal attitude. But such a scheme violates the complexity barrier and has not been attempted. Another solution is to rotate the whole engine mounted on the wing in a complete nacelle. The German XJ 101 proved this does work but the wing is heavy and rolling is sluggish; failure of one engine thrusting vertically far from the centre of gravity creates a hazardous situation. The next method is to rotate only the engine efflux: this is the solution adopted in the Harrier family of aircraft. The Rolls-Royce Pegasus engine is mounted in the fuselage under the high wing and compressed air and exhaust gases flow from the engine through four rotatable nozzles mounted along the fuselage sides. The pilot can rapidly direct these four parallel thrusting gas streams at any angle from directly aft to vertically downward.

In this way all the thrust is always available–it is changed in direction or 'vectored' at his command. Perhaps this is the secret of the Harrier's success for, unlike the other two methods described, it moves only small and simple parts of the aircraft (i.e. the four metal nozzles) instead of either whole engines or the whole aircraft (fig. 57).

Fig. 57 British Aerospace Harrier V/STOL jet lift fighter

Many other experimental aircraft used combinations of engines; usually one specialising in forward flight, the other being employed exclusively for take-off, thereafter being shut down. Such a solution appeared attractive since in the 1960s very high performance lift jets were made specially for this task. As they were only used for brief periods they could provide high thrust for low weight (T/W of 20 was achieved); as they did not need to be stressed for very high speed airflow through them in normal aircraft flight they were relatively smaller and lighter than cruise engines of the same thrust. The problems they brought, however, were severe. Usually mounted vertically they required their own air intakes on the top of the body and exhausts beneath. Air had to be persuaded to flow along the body and then turn sharply with minimum losses into the engine. After take-off the intake and exits had to be closed by airtight doors. Once aloft these engines were then a weight penalty and they occupied valuable space required for fuel and operational systems in the close-packed inside of a small military aircraft. In fact the Soviet YAK 36 employs this system. Lift jet aircraft can only rotate the lift engines a little, which restricts the acceleration possible along the flight path. The Harrier, however, can use its wing and flaps to take off from a runway like a conventional aircraft to carry heavier loads. It is for this reason that the Harrier's V/STOL title implies very great operational flexibility.

Other approaches to the problem of providing jet lift have sought to use jet energy indirectly, by driving fans or augmentors which offer a larger, slower airflow which does not heat or disturb the ground surface as much as does a pure jet. Fans can be driven either by a small lift jet, or by shaft power or by a tip drive in which exhaust gas is ducted to blow on to turbine blades fixed to the tips of the fan blades, beyond the fan shroud. Such tip-driven fans have been mounted inside wings but this leads to a complex airflow–the downward fan flow is in conflict with the wing circulation flow–and the large diameter fans occupy a considerable volume of wing space. The most ambitious augmented arrangement was the North American Rockwell XV-12A in which a supersonic engine exhaust flow could be ducted forward along the fuselage and then passed outward within the wings, eventually exhausting downwards through spanwise slots or hypercirculation nozzles. These were arranged to induce a much larger flow from outside and thereby increase the momentum of the exhaust. It was hoped to augment the primary jet thrust by an additional 80% by this means but the compressed air ducts were long and tortuous and suffered pressure losses so that such values were not in fact achieved. There were also many moving surfaces and doors over most of the wing surface which implies both heavy wing structure and leaky high drag surfaces for cruising flight, and the wing is essentially unable to carry internal fuel or external weapons.

Three kinds of jet lift system lend themselves to the different requirements of transport aircraft. Not such high demands are made on speed, performance or packaging density and solutions are quite different from those proved successful on military designs. The first is the multiple jet fan propulsion system mounted along both sides of an otherwise normal fuselage. These are used for vertical take-off and landing and are shut down in flight. Cruising flight propulsion comes from two or more normal fanjet engines mounted on wing pylons or at the tail. The second employs multiple gas turbines in body and nacelles driving large diameter fans in rotatable wing nacelles with a balancing fan in the fuselage nose. Cross shafting and gears give safety in the event of engine failure but the shaft power required is nearly an order of magnitude greater than that needed for a helicopter of the same weight. Third is a combination of two Pegasus vectored thrust engines mounted below the wing which also supply compressed air/exhaust to an augmentation arrangement mounted spanwise at the trailing edge of the wing. This 'augmentor-wing' concept developed over many years by Whittley at De Havilland Canada, uses the accelerating effect of the engine efflux to increase lifting circulation over the whole wing.

There are many special aerodynamic problems inherent in jet lift aircraft. In detail they vary considerably from type to type but will now be described briefly with reference to the Harrier which now has considerable operational experience from many kinds of environment.

The Pegasus engine gives a thrust of 21 500 lb (96.4 kN) at take off and swallows 420 lb (190 kg) of air each second. Since this is the volume of air contained in a typical house a large intake is needed to collect it. But if the intake area were too big it would limit forward flight speed by excessive drag. To overcome this conflict of requirements auxiliary inlets open automatically when high thrust is needed at low speed and close when not needed. Figures 58 and 59 show the very different streamlines in the two cases. Another vital

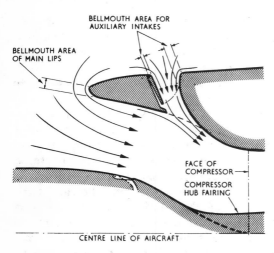

Fig. 58 Horizontal section through the intake, illustrating operation of the auxiliary intakes at VTOL conditions

feature of the Harrier intake aerodynamics which is not obvious is its short length. This was necessary to reduce the overall size and hence weight of the aircraft but it resulted in the airflow into the engine having to change

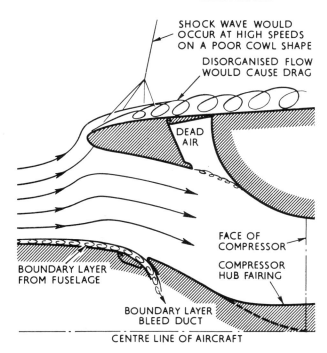

Fig. 59 Aerodynamics of the Harrier intake at high speed

direction and speed rapidly as it entered the two separate 'elephant-ear' intakes and then coalesce in front of the circular engine face. A device to remove the fuselage boundary layer by natural suction when travelling at high speed and low altitude was needed to persuade the air to maintain a steady streamline flow and not break away and cause eddies which would have reduced engine thrust.

Other aerodynamic effects are created by the high temperature, high velocity gas flows which, when vectored towards the ground impinge on it with force, and turn horizontally to form sheets of outward-flowing gas parts of which may be sucked into the intake. This must be avoided or minimised since the engine thrust is reduced if it swallows hot air. The jet impact on the ground may also throw up debris which could be sucked into the engine and damage it. Such a novel jet lift aircraft as the Harrier has to be operated differently from a conventional type and the techniques of aircraft design and correct piloting could only be really learnt from flight experience. Many model tests were made of all these aerodynamic features but could not be fully representative of the exact and very complicated mixed flows that are created during VTOL. Such matters become of increasing importance as more powerful versions of the Harrier are developed, especially for a new class which, although retaining the fundamental simplicity of the four vectoring nozzles, goes supersonic by virtue of additional thrust provided by burning extra fuel in the compressed air passing through the front two nozzles. This is known as Plenum Chamber Burning or PCB.

Two other special features of the Harrier remain to be discussed. Although provided originally for V/STOL the vectoring nozzles can be operated when in high speed flight as a means of augmenting the manoeuvrability when engaged in air combat. This trick is called VIFF (Vectoring In Forward Flight) and has permitted the Harrier to outmanoeuvre other more expensive adversaries not possessing this feature. The Harrier manages to keep balance when hovering in the air by means of compressed air reaction jet forces as did the Flying Bedstead. Compressed air outlets are provided at nose and tail for pitch and yaw and near the wing tips for roll control. This has subsequently permitted yet another special kind of flight behaviour–the Ski Jump. A ramp is mounted on the end of a ship or short ground strip so that after completing a STO the Harrier is pitched upwards to leave the deck at an angle of from 7° to 20°. It reaches free air at very much less than stalling speed but with upward velocity, and it keeps pointing in the right direction because of the reaction control jets. The high engine thrust accelerates the aircraft rapidly to full flying speed which is reached in level flight at about 200 ft (60 m). In effect, part of the take-off run is in the air! By this means much heavier payloads can be lifted off from much reduced take-off runs.

The history of jet lift aircraft is one embodying solutions to a wide variety of complicated mechanical and thermodynamic problems of which aerodynamics has played a vital role. The importance of careful investigation going hand in hand with good simple design could hardly be better emphasised than by this story.

Supersonic aircraft and missiles

Supersonic, in the strict sense of the word, means any speed greater than that of sound, but, as hypersonics begins at $M = 5$, only 'low' supersonic speed aircraft will be considered in this section. Mach 2 is 1320 mph (590 m/s) and Mach 3 is 1980 mph (885 m/s) in the stratosphere.

Although all the basic aerodynamic features of the subsonic aircraft are also found in the supersonic aeroplane, e.g. lift, vortices, boundary layers, body-wing interference and propulsion, the existence of shock and expansion waves introduce fundamental differences. Their wave drag can add about 50% to all other drags put together, the increase being most marked just above a Mach number of 1 as seen in fig. 60. The wave drag of a body or wing increases rapidly with its 'bluntness', hence supersonic bodies are long and pointed (lengths can exceed 20 diameters), and wings are pointed wedges whose maximum thickness may be as little as 1/30th of the chord. If the wing plan shape is swept back to lie behind the shock waves, the flow over the wing leading edge is subsonic and the drag rise is lessened. These considerations lead to aeroplane shapes such as those illustrated in fig. 68. This aeroplane also demonstrates the 'area rule' arrangement of body, wing and engine pod shapes, and their relative positioning in order to keep interference to a minimum at transonic speeds. The magnitude of the drag reduction achieved by this is also shown in fig. 60.

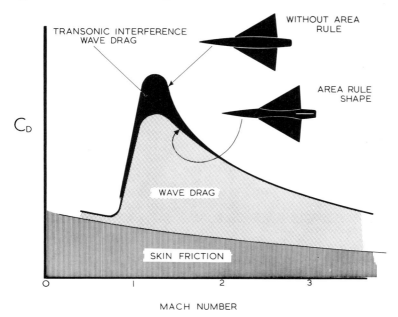

Fig. 60 Drag coefficient variation with Mach number

Two special configurations have been developed by aircraft designers in responding to the extra aerodynamic influences occurring in supersonic flight. The first is the swing wing which, although first advocated for transport aircraft, has in fact only been used for military fighters and bombers; the second is the slender wing Concorde–the first, and at the time of writing, the only technically successful supersonic airliner.

Swing wings (variable geometry)

The rationale of the swept wing in reducing transonic drag was described on page 134 and this has also been applied quite successfully on many supersonic designs, some even reaching over Mach 3. But swept wings have a disadvantage at take-off and landing, because the lift coefficient is markedly less than that for straight wings at given incidence. Furthermore, for any given structural length of wing, the span is reduced by sweep. Increasing the incidence is of limited value since it requires a long (and therefore heavy) undercarriage to avoid ground scrapes, and the drag is increased too.

Military aircraft often need to operate off short runways, and thus seek reduced take-off and landing speeds–ideally requiring straight wings of large span. Clearly, if the sweepback can be changed in flight, the wing can be set at the best sweep for any particular condition of flight–say fully swept to perhaps 65° at maximum speed, or for high-speed flight in gusty weather at low altitude (to reduce bumpiness of flight), and moderate sweep (45°) for high-g manoeuvring at, say, $M = 0.9$.

The first full engineering studies to solve all the practical difficulties were completed by a BAC team under the late Sir Barnes Wallis, who later introduced the scheme to the USA, where it was adopted for the F-111 fighter bomber. Current examples include the Panavia Tornado in Europe, the Grumman F-14 Tomcat fighter and NA Rockwell B-1 bomber, and several Russian aircraft. Of these, the Su-17 Fitter C is interesting in that it was developed from a conventional swept-wing aircraft. Others include the Flogger and Fencer fighters, and the Tu-22 Backfire bomber.

It is unfortunate that the term 'variable geometry' is used sometimes to describe the swing-wing, for geometry correctly means the shape in general– not only the wings. Air intakes and engine nozzles, for example, are other examples of variable geometry used on aircraft–as indeed are the many changes of shape seen when flaps and other high-lift devices are lowered for take-off or landing.

The Concorde[36]

Following the successful supersonic fighters of the late 1950s it was inevitable that attention should be turned to a supersonic airliner. As early as 1956 a Supersonic Transport Aircraft Committee was created in the UK from which eventually appeared the Anglo-French Concorde flying first in prototype form in March 1969. Although its beautiful and simple slender shape is now familiarly associated with this excitingly different regime of flight this was not a foregone conclusion in the late 1950s when British and French research on the problem began in real earnest. The story of the emergence of the new shape is worth telling briefly because it typifies the ways in which complex new aerodynamic behaviour is learned and the slow, painful and somewhat chaotic processes involved in aeronautical progress.

Many configurations were at first tried in attempts to project an aircraft to fly about one hundred passengers at a Mach number greater than 1.0. Those with small span, thin, unswept wings (like the F-104 fighter) had inadequate lift/drag to attain the desired range for a reasonable sized aircraft. A variant of the swept wing shape, the M-wing employed also forward sweep out to mid span and sweepback outboard. Canard foreplanes (fig. 61) instead of tailplanes did not improve lift/drag sufficiently and recourse was had to variable sweep designs. The Hawker Siddeley 1011 swept its high aspect ratio wings in combination with a high tail. As with the M-wing designs, by flying no more than $M = 1.15$ sonic boom shock waves were unlikely to reach the ground because of increasing air temperature at lower layers–the improvement in speed beyond that of the subsonic airliners would be worth having. There was concern however that the strong shock waves normally expected to

Fig. 61 The North American B-70 Mach 3 bomber. (Note the large vortices over the leading edges of the wing and the vortex wake behind the foreplane)

be absorbed by the atmosphere could be focussed during turns or in unusual weather conditions to give widespread damage on the ground. Interest eventually began to concentrate on the delta planform. This had already flown successfully on large and small subsonic aircraft and with a few supersonic military types, but by itself was still not aerodynamically efficient enough for the demanding task set by the designers.

Once again Küchemann was involved in a large research effort to find the solution but by now he was working at the Royal Aircraft Establishment in the UK leading British and French colleagues. He pointed out that for low lift-dependant drag, the lift needed to be spread over a certain *span*; for low supersonic wave drag the volume needs to be spread over a certain *length*, and for low supersonic lift-dependent drag the *lift* needs to be spread over a length. So he pointed to the entire class of wing planforms suitable for efficient long-range supersonic flight, in terms of the fraction of the characteristic rectangle occupied by the wing. The 'Gothic' planforms and the 'Ogee' planforms emerged as more interesting than the simple delta–with the latter finally winning the choice for Concorde, for its combination of good supersonic behaviour with good high-lift characteristics when landing or taking off.

Hundreds of models were tested and thousands of hours of wind tunnel time were used. Even once the Concorde shape was decided on six major models were needed in nine different French and British wind tunnels.

Much fundamental research into the airflow over such shapes had indicated a quite new type of lifting effect. Unlike the subsonic classical unswept wing which shed its circulation in vortices which collected from the wing tips (fig. 48) the slender shapes created a vortex flow all along the leading edge (fig. 62). This can be understood from the roughly similar type of flow experienced over flat roofs in chapter 5 and fig. 30.

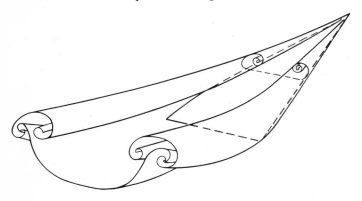

Fig. 62   Model of the flow past a lifting slender wing (after R. L. Maltby)

At high incidence (say greater than 10°) air flowing up to pass over the leading edge is drawn down into the large space in the aerodynamic 'shadow' of the wing and turns inwards and downwards in a large vortex flow. This will happen to the same extent on any wing leading edge swept-back more than about 45–50° but the Concorde ogee planform enhances the effect. The magnitude of this, at an incidence of 20° is significant: compared to a straight 60° wing the Concorde shape provides 113% more lift, of which 63% is accounted for by the leading edge vortex formation. Moreover this vortex flow does not break down until very high incidence is reached. This is why the Concorde approaches and lands at a high angle to reduce speed and is the reason for the drooped nose to improve pilot vision.

Drag is reduced (compared with a delta wing) in supersonic flight by as much as 40%. These gains, the result of an immense amount of theoretical and experimental work, following up intuitive judgement, at last gave a viable design.

The heat generated in the supersonic boundary layer creates several critical problems unknown in the subsonic aeroplane. At Mach 2.2, cruising at 60 000 ft (18.3 km) the skin temperature could be 170°C, rising to 400°C at Mach 3.2. Although light alloy structures suffice for $M = 2.2$, steel is essential for the higher speed. Special fabrication techniques are necessary to maintain a smooth exterior and to cope with the transient heating stresses. Other problems are presented by the abnormal heating of windows, tyres, fuel and passengers.

The fuel volume could be held in the large area 2% thick wing for an acceptable structural weight and the four engines, in double nacelles each side under the wing did not interfere with the important vortex flow over the top surface. Apart from the overall aerodynamic lift and drag efficiency the airflow was also well behaved without buffet or vibration and the Concorde not only handled familiarly and smoothly at supersonic speeds but at all other speeds as well. Care was needed to avoid rolling instability caused by an over wing vortex bursting on one side before it had passed over the trailing edge. These and many other important aerodynamic issues led to many refinements to the configuration such as the heavily down-cambered leading edge, the variation of twist across the span and the appearance of minor strakes just below the pilot's windows. These prevent asymmetric shedding of vortices from the fuselage in sideslip when at high incidence. During a period of six years the combined effect of all these and other changes are impressive: 25% reduction of C_D, 25% increase in trimmed C_L and 6% favourable shift of aerodynamic centre.

In the USA, many different concepts were worked on intensively without success. At first, much higher speeds were planned–first Mach 3 and then 2.5, but these demanded titanium structure of enormous expense and high weight, and very costly cooling systems. A Swing-wing design was tried, but eventually the last US project to die amounted to a design very like Concorde, but larger. The larger size would have given somewhat cheaper running costs per seat, but more airport noise and larger launching costs, and after spending the equivalent of about £600m the US SST was dropped.

The Soviet Tu-144 supersonic transport (which has been through very major redesigns, and superficially looks rather like Concorde) has not been so well thought out–not only basic concepts have to be right but all the detail refinements are essential. In particular, the bypass engine chosen had high fuel consumption at supersonic speeds, and the external detailed shaping was crude.

Apart from the location of the engine nacelles beneath the wing, set back so the inflow will not influence the wing leading edge, the air intake design had to satisfy many different aerodynamic requirements. The intake slows down the supersonic flow smoothly so that it enters the engine face at subsonic speed. To do this it has adjustable ramps to reduce the internal area progressively, and elaborate boundary layer slot suction arrangements (fig. 63). At take-off it needs to be fully open, aided by auxiliary inlets. It has to respond quickly enough to sudden demands for more engine power and more particularly to an engine failure which would block the flow and spill the excess over into the adjacent engine. Considerable model testing followed by full scale testing of the complete engine nacelle under a Vulcan bomber finally gave a satisfactory system which also needed a very reliable and sophisticated electronic control system to handle all the changes and adjustments automatically.

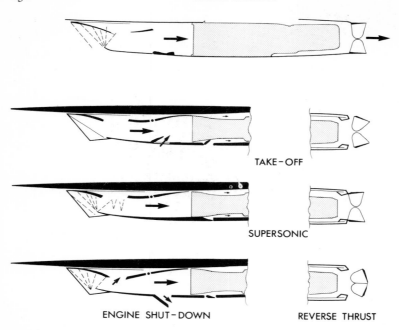

Fig. 63   Variable intake and nozzles of the Concorde

In spite of the relatively small increase in altitude above that of present jet transports (40 000 ft (12.2 km)) three new phenomena are expected: a high ozone concentration (which could be toxic to passengers), some increase in cosmic radiation, and abrasive damage to the structure caused by high-speed impact with minute ice crystals. Ozone entry into the cabin from the cabin conditioning system is eliminated by two means; the air is taken from the engine compressors at high temperature which breaks down the ozone and, as a further safeguard, a catalytic filter is fitted. The other ozone problem raised in connection with supersonic flight is the possible reduction of the Earth's natural ozone layer by nitrogen oxide emissions from the engine exhausts. This has been calculated to be very small compared with other influences such as soil bacteria and aerosols. The increase of ultra-violet and cosmic radiation at sea level due to reduced ozone barrier layer and a possible consequential increase of skin cancer is estimated to amount to an equivalent ninety seconds of sunbathing on the beach.

Another serious worry about supersonic flight is the destructiveness of shock waves as they sweep out of a path below and behind the aeroplane. During the climb, peak pressure jumps on the ground could reach $2\frac{1}{2}$ lb/ft^2 (120 Pa) spread over an area of roughly 2 by 10 miles (3 by 16 km). A Mach 2 USA record breaker (the B-58 shown in fig. 68) broke windows and damaged buildings along an 800-mile (1300 km) path. The operation of supersonic airliners will undoubtedly be restricted, geographically, by this aerodynamic phenomenon.

Research and design studies have indicated ways in which the SST could be improved further; quieter engines, lighter structural materials, larger size, greater productivity, etc. but the unfavourable economic climate of the late 1970s and early 1980s goes against the large financial investment for such new projects. An unexpected effect of the oil crisis has been a vigorous examination of liquid hydrogen as a substitute for kerosene as aviation fuel. Design studies show that an SST using liquid hydrogen could be only half the weight of its kerosene-fuelled equivalent but of larger volume leading to reductions of noise and cost. Perhaps the SST will return to service again after the year 2000 in a new guise using a new fuel.

Supersonic aerodynamic missile shapes are either the interceptors with cylinder-cone bodies and small square-shaped wings, or the canard delta shapes of the longer range, cruise bombing weapons. Some defence missiles have special aerodynamic problems because they have to manoeuvre at very high incidence (60°) when vortices generated from the upper surface of the body distort the airflow and de-stabilise the controls.

Hypersonic flight (M > 5)

The US X-15 (fig. 9) is the only aeroplane to achieve sustained manned flight at a Mach number exceeding 5 (fig. 67). As aeroplane speeds increase, so does the altitude and also the aerodynamic heating, as shown in Table 12 calculated for a typical shape at 5° incidence.

Table 12

Regime	Airspeed (ft/s)	(m/s)	Mach number	Altitude (ft)	(km)	Relative air density	Mean body equilibrium temperature (K)
Supersonic	2 000	610	2.1	57 500	17.5	0.107	370
	4 000	1 220	4.1	87 500	26.7	0.025	735
Hypersonic	8 000	2 440	7.8	116 000	35.4	0.006	935
	16 000	4 880	13.3	156 000	47.5	0.0012	1 100
	20 000	6 100	16.3	182 000	55.5	0.00047	1 150

In calculating the mean body temperature, the real gas effects described in chapter 4 have been included. In spite of the heat energy absorbed by the molecules and radiated from the surface, the hypersonic flight skin temperatures are too high for steel. Some aspects of the heating problem are similar to those of the ballistic missile (chapter 8) but, whereas the latter has predominantly a rapid rate of heating, a hypersonic aeroplane must fly for a considerable time at the temperatures indicated in the table above. The problem has been tackled from two standpoints: the aerodynamic solution, by introducing new shapes, giving reduced heat transfer; and the metallurgical solution, by adopting radically different high-temperature structural metals.

Hypersonic wing leading edges can be quite blunt (e.g. a radius of 2 in (51 mm)) thus giving a strong shock well ahead of the leading edge. Behind this is a zone of subsonic continuum flow and a laminar boundary layer which alleviates the heat transfer in this critical region. To reduce the drag of the rounded leading edge this is markedly swept back to as much as 80°. In addition to these radical changes in shape, hypersonic vehicles need either new structural metals such as molybdenum, niobium or beryllium, or the removal of heat by cooling the skins with gas or liquid. One method is to pass liquid hydrogen beneath the skin in a jacket so that it absorbs heat and can be used subsequently as a rocket fuel. Alternatively, in transpiration cooling, water is bled through a porous surface into the boundary layer where it evaporates and thus cools by diffusion and latent heat.

Two types of aeroplane shapes which were investigated for hypersonic flight are shown in fig. 64. The first has very thin, highly swept-back wings

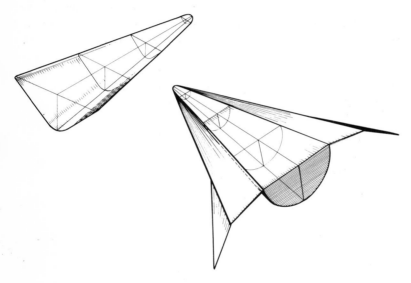

Fig. 64 Hypersonic aeroplane shapes

with a semi-circular body beneath. The idea behind this is that the body shock wave creates a region of increased pressure which passes across the underside of the wing, thus increasing its lift, while the downturned wing tips further deflect any side flow downwards. The lift-drag ratio may exceed 6. The other shape is wingless, and in exchange for a lower lift-drag ratio ($2\frac{1}{2}$) offers a greater volume for propellant or payload.

Such vehicles are boost-gliders, that is, they are rocket-boosted to a maximum speed which is reached when all propellant is burnt, and thereafter glide with engines off. The distances travelled by such vehicles are given in Table 13.

Table 13

Maximum Mach number	Range in miles (km)	
	$L/D = 6$	$L/D = 2.5$
3	280 (450)	130 (209)
5	600 (965)	270 (434)
10	2200 (3540)	950 (1530)
15	4700 (7560)	2040 (3280)

Air-breathing propulsion for hypersonic vehicles may eventually become possible as an outcome of much research into and development of the ramjet. At a Mach number of 7 the ramjet's overall thermal efficiency could be double that of the turbojet. An interesting proposal is external burning of fuel on the under-surface of the hypersonic aeroplane so that the exhaust expansion gives part thrust and part lift (fig. 65).

Even at very high airspeeds the airflow pattern is composed of different regions as in the simple picture of fig. 3. Figure 66 shows a typical flow pattern around supersonic combustion in which all the flow patterns are to be seen.

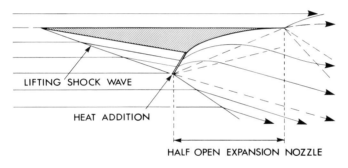

Fig. 65 Scheme of a propulsive lifting body

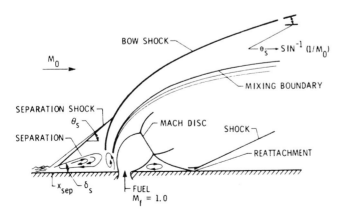

Fig. 66 Aerodynamic features of the transverse jet interaction

AERODYNAMICS

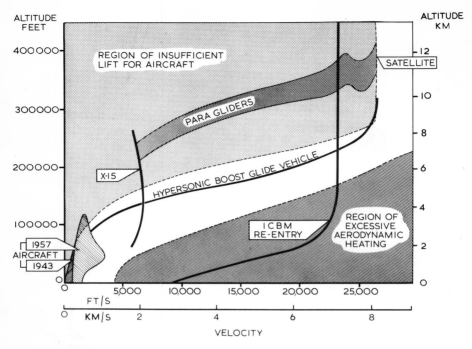

Fig. 67 Flight corridor for aircraft, rockets and space vehicles

An indication of the rapid extension of aeronautics into new regimes of flight is given by fig. 67. The white 'flight corridor' is a performance region in which supersonic and hypersonic aeroplanes could operate. For a given airspeed a vehicle cannot fly below a certain altitude because the heating is too great, nor fly too high where the air density is too low to give enough lift. The corridor chart also links aeronautics with astronautics, and this subject is discussed in the next chapter.

8
Aerodynamics in Space

He rode upon the cherubins, and did fly: he came flying upon the wings of the wind

Prayer Book

In this chapter we consider motions in the atmosphere above aeroplane heights, i.e. above 30 miles (50 km). The region of significance to re-entering spacecraft extends up to 200 miles (300 km) and beyond this the atmosphere merges imperceptibly with the plasma of space. Magnetohydrodynamic effects are of overwhelming importance in these regions, and they belong to the subject of cosmic electrodynamics.

The rarefied atmosphere from 30 to 1000 miles up (50–1600 km)

Some characteristics of the upper atmosphere are shown in Table 14. The different levels have been given names, e.g. stratosphere (uniform temperature), ionosphere (predominance of positive ions and free electrons), and exosphere (where atoms readily escape the gravity pull of the earth). The large increase of free electrons above 250 000 ft (76 km) and the reduced air mass density give rise to high electrical conductivity; indeed electrical currents have been detected, some consisting of millions of amperes. The high electron concentration also affects the air drag of moving objects above about 250 miles (400 km).

A great many features of this region still remain to be discovered but the co-operative world scientific programme of rocket experiments is providing an increasing fund of knowledge. For example, in autumn 1961, a probing rocket discovered, for the first time, a belt of helium gas extending from 700 to 2000 miles (1100–3200 km) above the earth, which has probably evolved from radioactive decay of minerals in the terrestrial crust.

Aerodynamics of spaceflight

Aeronautics is concerned with the carriage of objects from one part of the Earth to another through the atmosphere. Military needs demand offensive vehicles which are either undetectable or difficult to intercept and, starting with the German V2, a large number of very fast ballistic rocket weapons have now been built. A 5000-mile (8000 km) range intercontinental ballistic missile (ICBM) reaches a maximum speed of about 15 000 mph (6–7 km/s) a maximum height of 700 miles (1100 km) and returns to Earth along a path inclined at about 30° to the horizontal. The US Atlas ICBM, also used for spacecraft launching, is shown in fig. 68. Spacecraft, which have become

Fig. 68 B.58 supersonic aeroplane and Atlas ICBM

possible as a by-product of guided missile technology, have many aerodynamic characteristics similar to those of the ICBM, but these are quite unlike those of the aeroplane. It might be thought that in the near-vacuum of space there would be no place for aerodynamics, but this is not so, for at very high speeds (18 000 mph (8 km/s)) even the very rarefied gas produces heating, pressures and forces that can be measured and which influence the behaviour of spacecraft. As in the lower reaches of the atmosphere, there are 'ballistic' and 'aerodynamic' craft, having quite different aerodynamic features, but because the 'air' in these regions is a complex mixture of electrified particles, electric currents and radiations, some of the concepts such as boundary layers and air temperature have to be interpreted rather differently. Laws of aerodynamic similarity still apply, however, and laboratory experiments can represent the fast flows of electrified gas past objects.

Aerodynamics of launching

As a large ballistic rocket accelerates upwards it passes through air of decreasing density and the dynamic air pressure $\frac{1}{2}\rho V^2 (q)$ reaches a maximum, usually near the beginning of the stratosphere. Thereafter, although the speed is increasingly hypersonic, the aerodynamic drag and heat transfer do

AERODYNAMICS IN SPACE

Table 14 Some characteristics of the upper layers of the Earth's atmosphere (*These are mean values. There are wide variations especially at high altitudes*)

Zo	Altitude ft (000)	Altitude km (000)	Kinetic temperature K	Density Slugs/cu ft	Density kg/m³	Mean molecular weight	Mean free path ft	Mean free path m	Collision frequency per sec	Free electron density per c c	Moving objects	Other phenomena
Troposphere	0	0	288	2.38×10^{-3}	1.23	28.9	2×10^{-7}	6.1×10^{-8}		0	Cars, ships	Natural
Stratosphere	35	56	220	7.4×10^{-4}	0.38	28.9	7×10^{-7}	2.1×10^{-7}	1.9×10^9	0	Aeroplanes	Aerodynamics
Stratosphere	150	241	278	3.6×10^{-6}	1.86×10^{-3}	28.9	1.5×10^{-4}	4.6×10^{-5}	1.0×10^7			
Ionosphere	250	403	183	7.8×10^{-8}	4.02×10^{-5}	28.9	6.7×10^{-3}	2.0×10^{-3}	1.8×10^5	1 (10^3 Night)		
Ionosphere	500	805	1070	3.0×10^{-12}	1.55×10^{-9}	28	1.6×10^2	4.9×10^{-3}	1.8×10	Day 10^5 10^4	Re-entering	Aurorae
Ionosphere	1000	1610	1420	8.5×10^{-14}	4.38×10^{-11}	21.9	4.6×10^3	1.4×10^{-3}	8.4×10^{-1}	10^6 10^5	ICBM space-craft Meteors	
Ionosphere	2300	3703	Temp. varies with time of day and solar activity in range 1100–2100	4.4×10^{-16}	2.27×10^{-13}	16	7×10^5	2.1×10^5	7.2×10^{-3}	10^5	Long lifetime satellites	
Exosphere	3700	5957		10^{-17}	5.15×10^{-15}	16	2×10^7	6.1×10^6	10^{-4}	10^5		Helium belt
Deep Space			3000			2				1	Spacecraft Comets Asteroids	Van Allen Radiation belts Cosmic electrodynamics

not become excessive. The drag losses during the atmospheric path represent only a few percent of the kinetic energy acquired. Rockets often have extremely thin steel skins ($\approx$ 0.015 in (0.4 mm)) and high airspeed tends to set up a panel flutter leading to complete structural failure. Launching vehicles are usually steered and stabilised by automatic tilting of the rocket motors, but they encounter disturbing lateral aerodynamic forces in turbulence or when passing through layers of the atmosphere of varying wind velocity.

Aerodynamics of objects in earth satellite orbits

The magnitude of the aerodynamic drag in space determines the lifetime in orbit of an earth satellite. The descent of heavy objects is primarily influenced by the aerodynamic drag-weight ratio, e.g.

$$\text{Drag/mass} = \tfrac{1}{2}\rho S V^2 C_D g/W$$

Apart from the drag arising from the change of momentum in encountering the mass of the gas, there are three other forces which are important for small or low density objects. These are:

(a) Solar radiation pressure ('solar wind'), which amounts near the earth to a steady stream of about 2×10^{-7} lb/ft² (9.6 μPa) with bursts (or 'gales') reaching a value eighty times as great.
(b) Electromagnetic or Lorentz force, arising from the interaction of the body's electric charge and the Earth's magnetic field. It acts perpendicularly to the velocity vector of the body, and is therefore similar to a 'lift' force.
(c) Electrostatic drag (Coulomb) force, acting between the body charge and the electrified gas of space. This is akin to the true drag, and leads to a loss of kinetic energy.

Evidence of these forces is revealed by accurate observation of the orbit changes of satellites. It was found that the air density varied with latitude and also with time over a period of about 28 days. This was a direct result of charged particles emanating from the surface of the Sun which rotates in this period. Nuclear explosions at great altitudes are also believed to cause marked density changes. The solar wind ((a) above) had a pronounced effect on the Echo balloon (100 ft diameter (30 m), 60 lb mass (27 kg)) by displacing its initial orbit from a perigee of 941 miles (1514 km) in August 1960 to 717 miles (1153 km) in February 1962, the corresponding values of apogee changing from 1052 miles to 1164 (1693 km to 1873 km). These forces also determine the motion of interplanetary dust particles near the Earth and would have affected the small metallic filaments of the West Ford radio communication-by-reflection experiment. Uncertainty in calculating the lifetime of these 'needles' gave cause for concern because of their possible interference with optical and radio astronomy. There is a great deal still to be

Space vehicle internal heating and cooling

An earth satellite will normally have a variable heat input depending on whether it is in the direct rays of the sun or in the Earth's shadow. Surface temperatures are usually within the range ±100°C, but the exact value depends on the colour and other details of the surface. The internal power supplies, electronic equipment and the human occupants have to be maintained within permissible temperature limits. If the cabin contains air (or oxygen-helium mixture), normal convection will not occur in the weightless state, and forced convection must therefore be introduced by fans. In an instrumented, evacuated space probe, heat transfer between the outer surface and the electronic equipment has been considerably influenced by outgassing of occluded air from equipment, and, incidentally, the leaking out of this gas through the satellite can alter the aerodynamic resistance of the atmospheric gases. Special environmental chambers are now used to expose prototype space vehicles to the high vacuum, solar energy and magnetic fields of space.

Aerodynamics of re-entering the Earth's atmosphere[37]

The re-entry of ballistic missiles, manned space capsules and meteorites involves similar aerodynamic phenomena but varying in degree according to size, weight, speed and entry angle. The general scale of events is described here for the steep entry of an ICBM.

The re-entry phase begins at about 350 000 ft (107 km) (relative air density 10^{-7}), when the aerodynamic pressures will begin to rotate the body if it is not aligned with the direction of motion. About 20 seconds later, at 250 000 ft (76 km) aerodynamic heat transfer increases markedly, reaching a maximum value of about 4000 hp/ft^2 (32 MW/m^2) with surface temperatures exceeding 2000°C for some seconds. The drag force reaches a maximum at a height of approximately 200 000 ft (61 km) giving a deceleration of about 40g. About 1 min after the first aerodynamic effects become noticeable, the re-entry body will have reached the troposphere (fig. 67). If the re-entry were made vertically downwards, the deceleration could reach 250 g.

A typical airflow pattern round a blunt re-entry body is shown in fig. 69. Even in this extreme condition of flight the basic flow patterns are evident–shock waves, the two boundary layers, and the wake–but the temperatures are so high that the air becomes ionised. The boundary layer therefore becomes conducting and it is difficult to pass radio transmissions through it. The wake remains ionised for some distance behind the body until the energy is dispersed into the surrounding gas, and it is from such wakes that radar signals are reflected.

The major aerodynamic problem in the steep re-entry of a ballistic missile is to minimise the heat transmission to the body. Whereas in the supersonic aeroplane the objective is to reduce wave drag (achieved by designing them fine and pointed in shape), in the ballistic missile the opposite principle applies, because by increasing the strength of the shock waves the energy imparted to the air is increased while the energy convected to the vehicle is reduced. Thus, in the first ballistic missiles which were extremely blunt, as seen in fig. 69, of the total kinetic dissipated 99% went into shock waves,

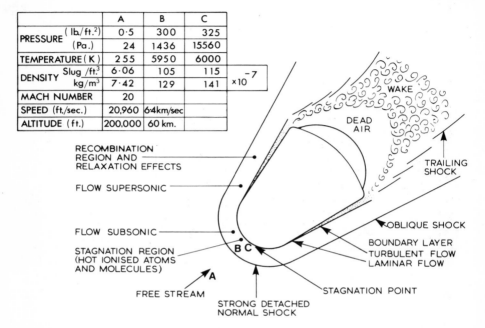

		A	B	C	
PRESSURE	(lb./ft.^{2})	0·5	300	325	
	(Pa.)	24	1436	15560	
TEMPERATURE (K)		255	5950	6000	
DENSITY	Slug /ft.^{3}	6·06	105	115	$\times 10^{-7}$
	kg/m^3	7·42	129	141	
MACH NUMBER		20			
SPEED (ft./sec.)		20,960	6·4km/sec		
ALTITUDE (ft.)		200,000	60 km.		

Fig. 69 Re-entry body airflow

radiation, and the wake, and only 1% into the body by heat transfer. This heat was absorbed by a heat shield, an inch or two thick, of copper, beryllium or graphite. The high drag of the blunt shape had the great disadvantage of making the speed below 100 000 ft (30 km) relatively slow, and thus increasing the chance of interception. It was overcome in the second generation of re-entry missiles which were of a cone-cylinder-flare shape, but they, in turn, had a correspondingly higher heat transfer. It could, however, be dissipated by the ablation of a coating of plastic material that absorbs heat energy in a very complex process. First, the material absorbs heat to raise its temperature, and then it begins to melt (or evaporate) and mixes with the boundary layer. Some materials melt into a viscous glassy substance and energy is absorbed in moving this downstream. Other materials, of a reinforced plastic, char and leave a highly insulating layer which reduces heat transfer in the later stages. Ablation technique has proved very successful and plastics are a great deal better than the metal heat shields (for the same weight) in preventing aerodynamic heating from reaching the vehicle itself. It has been reported that the Russians have used the hard wood, *lignum vitae*, as a re-entry heat shield, but there has not as yet been any scientific evidence to confirm this.

Manned space vehicles

Three different kinds will be described, viz. the Mercury pure ballistic capsule with a single astronaut, the Apollo three-man lunar spacecraft and the Space Shuttle.

Mercury ballistic capsule (fig. 70)

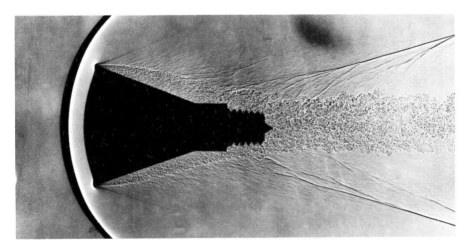

Fig. 70 Manned ballistic capsule

The aerodynamic problem of manned re-entry vehicles is to reduce the deceleration and heating to tolerable values. Evidence of meteorite behaviour would, at first sight, suggest this to be impossible, but when the aerodynamic forces and heat transfers can be calculated, it is possible to find acceptable solutions by choosing the correct shape, size, weight and speed. A ballistic technique was adopted for the first Russian and American man-in-space flights, but there are several other solutions.

The Mercury capsule embodies principles similar to those of the ballistic missile but the re-entry path descends more gently–at about 1° below the horizontal in comparison with the 30° of the ICBM. The maximum deceleration is thus reduced to only 8 g. Since any ballistic vehicle has to be 'aimed', the arrival of the Mercury capsule within a recovery zone in the ocean demands a precise pre-calculation of a large number of effects prior to initiating the downward descent by rocket deceleration. The atmospheric rotation displaces the arrival point by 150 miles (240 km) and the oblate shape of the earth, if not allowed for, could give an error of 300 miles (483 km). Terminal errors up to 200 miles (322 km) have, in fact, occurred.

Apollo

Essentially developed from the Mercury shape this larger craft (154 in (3.9 m) diameter) had to re-enter the atmosphere at the much higher speed of 25 000 mph (11.2 km/s). Its centre of gravity was located off-centre so that it flew at an angle of incidence of about 20° to the flight path and hence developed a lift force which could be used to bend the re-entry trajectory and correct for any displacement errors. It is not self-evident that such a blunt conical shape without any tail fin surfaces would be stable and a very extensive series of wind tunnel tests measured the static and dynamic stability characteristics over Mach numbers from 0.2 to 18.7 in sixteen tunnels. The testing covered a very wide incidence range from flat face forward round to apex forward since it was imperative that the heat protection shield on the base should face the main airflow. With apex forward there was a flow separation between 10 and 50° which in fact created an undesirable region of stability, hence the re-entering spacecraft had to be prevented from over-swinging to this condition after re-entry. The effects of various protuberances such as aerials, rocket vents and cavities were shown to be negligible. The heat shield consisted of a fibreglass honeycomb matrix bonded to a stainless steel substructure subsequently filled with the ablative material AVCOAT 5026–39. Its thickness varied from 1.71 in (4.3 cm) near the front corner to only 0.33 in (8 mm) near the apex on the downwind side. Eighty per cent of the heat load was taken by the front face (reaching a temperature of 5000°F (3030 K)), 17% by the windward side of the cone and the remaining 3% by the leeward side of the cone. The ablation material prevented the bonded steel from exceeding 600°F (589 K).

A special aerodynamic problem was to calculate the electron density created in the flow past the capsule (the entry plasma sheath) which prevented the transmission of radio messages from the Apollo to Earth during re-entry. Firstly streamlines were calculated at an angle of incidence of 20° at an airspeed of 34 000 ft/s (10.4 km/s) and a height of 200 000 ft (61 km). The chemical kinetic relationships of aerothermochemistry were adapted to allow for rarefied gas effects at high altitude and included electrified species of nitrogen, oxygen and nitric oxides.

Rays were traced through the high electron density layers of the flow within the main shock wave to determine the attenuation of the radio transmission at a frequency of 2.287 GHz. It was accepted that little could be done to alter the predicted characteristics (although 'seeding' with a metallic power was considered), hence the results were used to predict over which part of the re-entry trajectory the radio 'black-out' would occur. Agreement with flight data was remarkably good.

Two other 'aerodynamic' features were the model testing in vacuum chambers of the paths of moon dust thrown up by the rocket jets used to perform soft landings or take off from the moon surface and experimentally establishing the shock wave and boundary layer flow created by a small attitude control rocket motor operating close to the spacecraft surface in the vacuum of space.

Space Shuttle

The Space Shuttle is a unique hybrid–part aircraft and part spacecraft–with some intriguing aerodynamic problems. A major objection to the conventional chemical rocket propelled spacecraft is the total loss of nearly all the expensive craft after launching with the return to earth of only a small percentage of the original mass. The aim of the Space Shuttle is to maximise the amount of spacecraft that returns to earth and which can be used again. Although some of the earliest project concepts for such systems proposed during the 1960s envisaged multi-staged winged craft that would completely return and be almost immediately reusable, various practical limitations including otherwise over costly development programmes have left the Space Shuttle achieving somewhat less than this idealised goal. The main returning component (the Orbiter) is a large big-bodied delta wing hypersonic rocket aircraft 120 ft (37 m) long weighing 96.5 tons (97.5 Mg) which is boosted into orbit by its own motors (1.12 million lb thrust (5 MN)), supplemented by two vast solid propellant rockets each providing an extra 2.65 million tons (26.6 GN) thrust. 680 tons (690 Mg) of additional liquid propellants for use by the Orbiter's motors are carried in a large expendable fuel tank. The two solid rockets return to the ocean by parachute for subsequent recovery and re-use.

The Orbiter design follows the principles for hypersonic aircraft. The wing has highly swept, well rounded leading edges and the excess aerodynamic heating is absorbed by insulating surfaces some of which need replacement after each flight. High incidence flight is employed to distribute the re-entry heat over the very large under surface area which is consequently exposed to high temperatures for a longer duration than, for example, the Mercury ballistic capsule. Arising from re-entering at a high incidence many surfaces only receive a moderate heat input and four quite different thermal protection systems are used. The most severe heating is along the leading edges and the nose of the fuselage–here carbon-fibre cloth is used which is impregnated with additional carbon and coated with silicon carbide (fig. 71). It protects to a temperature of 1650°C. All the undersurface and around the cabin windows is protected by reusable high temperature surface insulation to 1260°C. Low temperature re-usable insulation is used for fins, part of the fuselage and parts of the top wing surface (648°C). Both insulants are borosilicate glass fibre but the high temperature kind has an extra coating of black silicon carbide. The remainder only requiring protection to a temperature of 398°C is a silicon-coated nylon. Extensive wind tunnel testing was needed firstly to establish the exact shape of the Orbiter and subsequently evaluate the heat transfer from the boundary layer. Heating commences at a very high altitude where the

Fig. 71 The underside of the Orbiter glows with re-entry heat

atmosphere behaves as a rarefied gas and aerothermochemical calculations are essential. The thermal protection is applied in the form of 31 000 tiles bonded on to the inner structural surface. Apart from the obviously important issue of aerodynamic heat input the Space Shuttle required the solution to many aerodynamic problems. The combined shape of Orbiter tank and rockets creates major interference airflows and shock wave interference especially near the speed of sound (fig. 72). The accelerating force of the rocket thrust is more than adequate to deal with the marked increase in drag that this causes but the pitching moment changes have to be corrected by rocket deflection quickly enough to keep the Shuttle on course. Although the take off is made vertically (as with a ballistic rocket) and hence aerodynamic forces soon become negligible, the Orbiter after re-entry has to position itself to land on a long runway. It therefore needs crisp roll and pitch control as does a normal airliner, but since it cannot open up its engines to go round again if the landing is inaccurate, certainty in the control of the approach and subsequent rounding out of the steep glide (started at 1700 ft (520 m)) has to be 100% sure. The landing behaviour of this unusual aircraft shape was first tested on a series of small experimental gliders taken aloft under large B52 jet bombers. The astronauts undergo many hours of practice landings in simulators which are fed with synthetic aerodynamic data based on calculation and

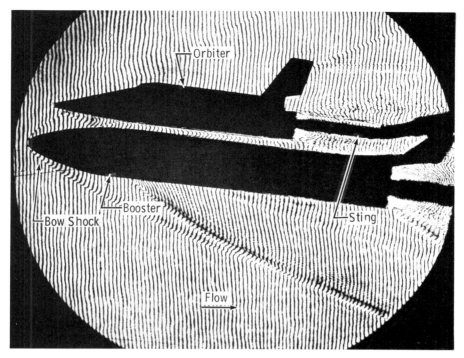

Fig. 72 Finite fringe holographic interferogram–Space Shuttle testing at Mach 8

experiment which reproduced the expected dynamic behaviour of the Space Shuttle many months before they had to fly it in earnest.

No other aircraft has been designed to fly up to Mach 23 and this must be attained on its very first flight. A typical trajectory would be:

Table 15

Time		Mach number	Altitude	
min	s			km
0	52	1	24 000 ft	7.3
1	53	3	120 000 ft	36.5
2	15	4	136 000 ft	41.5
3	50	6	61 n mile	113
6	30	15	70 n mile	130
8	30	23	63 n mile	117

Considering the need for safety and to ensure all systems functioned together, the delays in reaching the first flight date seem understandable.

A serious noise problem was cured in a novel way. The rocket efflux creates a very high noise, potentially 168 dB compared to the 160 dB satisfactorily withstood by Saturn V in launching Apollo. However, the sensitive instruments in the payload bay of the Orbiter are very much closer to the source of the noise and it was feared would be damaged during the first few seconds of take off as the noise was reflected back from the metal platforms at the launch area. The noise was reduced to a level of 162–3 dB by the creation of artificial raindrops from quench nozzles which dumped 300 000 US gallons (1136 m^3) into the space into which the rockets exhausted.

The possible effects of the rocket exhausts on the ozone layer have been carefully investigated. H. S. Pergament developed an aerodynamic computer program called the atmospheric interaction plasma program which evaluates several reaction mechanisms between the combustion products in the exhaust plasma and downstream shock waves and turbulent mixing zone of the external high altitude air (fig. 56). The thirteen exhaust gases contain various species of oxygen, hydrogen, nitrogen and chlorine, nitric oxide and hydrochloric acid. Chemical kinetic reaction rates were developed and calculations compared over both full size and small experimental models to check for similarity relationships. It was concluded that the level of nitric oxide produced in the stratosphere by the Space Shuttle would be only 1/500 that of hydrochloric acid.

The Space Shuttle is the world's first truly orbital aircraft and is expected to develop until the end of the century performing a variety of orbital missions and launching other rocket craft into deep space. It is interesting to speculate if this is the 'ultimate aircraft' or whether it would be superseded. During the 1960s various alternative means of providing recoverable space launchers were proposed, some of which employed air-breathing turbines and ramjets. Theoretically they could be lighter and take off and land horizontally. The aerodynamics of the intakes to feed mixed engines over such a wide Mach number range are undoubtedly formidable, as are also those of surface cooling. By A.D. 2000 the issue will probably have been decided–probably on cost grounds.

Inflatables and paragliders

The main factor determining the severity of deceleration and heating during re-entry turns out to be W/C_DS (W is the weight, C_D the drag coefficient and S the surface area). Typical values of this quantity are about 100 lb/ft^2 (4.79 kPa) for Sputnik II and 10 lb/ft^2 (479 Pa) for the 21-in (0.53 m) diameter Vanguard satellite, which would both vaporise completely during re-entry. However, if W/C_DS could be reduced to 2 lb/ft^2 (96 Pa) or less, the temperature need not exceed 1000°C. This fact has led to some enterprising research into alternative vehicles which do not employ the rigid metal bodies of the earliest satellites. An inflatable, lifting vehicle has been proposed in the Dyna-Soar programme, and another scheme is based on the opening of a large steel umbrella. One novel solution is the 'paraglider', which is a paper-dart shaped, metallised fabric 'kite' with leading edge booms, which can be folded away inside the launching rocket during ascent and deployed for re-entry. Its wing loading with a 1500 lb (680 kg) payload can be as little as $\frac{1}{2}$ lb/ft² (24 Pa) and it glides at altitudes much higher than hypersonic aeroplanes and with lower surface temperatures than a purely ballistic device. These unusual devices emphasise the virtue of an open-minded approach to the unfamiliar problems of flight in space. The next ten years should show some surprising developments, and already man-carrying, low speed, inflatable and paraglider aeroplanes have flown.

Re-entry of radioisotope power source

Long distance spacecraft carry radioisotope power sources such as SNAP-19 fitted to Pioneer probes. To cater for a possible rocket failure prior to achieving the desired injection speed it was necessary to prove that the fuel cask could survive the impact and heating of re-entry in spite of the rest of the craft disintegrating and burning up. The cask was of unusual shape, viz. a hexagonal prism with flat ends so that heat transfer rate could not be predicted with any certainty. Accordingly a 1/2 scale model was tested in a Mach 6 wind tunnel at several combinations of angle of attack (up to 90°) combined with angles of roll. Schlieren pictures of the complicated flow patterns were taken which showed where vorticity was created from corners which locally increased heat transfer coefficient. An approximate theoretical method was shown to give an overestimate compared to measured values. This simple example shows the great diversity of aerodynamic flows encountered in applications and how new products create problems unsolvable by existing theories. It is unlikely that this situation will ever change.

Other re-entry aerodynamics

A great deal of theoretical and experimental aerodynamics has been devoted to the re-entry problem, and this data has been collated and applied in very different ways. Two examples are given which emphasise the imaginative scope now offered to research workers engaged in this type of work.

Entry into other planetary atmospheres

Astrophysics has provided a variety of data on the constituents of the planetary atmospheres, based mainly on spectrographic, optical and radio surveys, and analogy with meteorological phenomena found on Earth. To make an exact estimate of the aerodynamics of a spacecraft entering an unfamiliar planet's atmosphere, however, far more information than is at present available will be needed. Pressure, density, speed of sound, viscosity and many other factors are vitally important, and some phenomena not found on Earth will have to be expected. But the situation has improved now instrumented space probes have passed through the upper layers of Martian and Venusian atmospheres, although the instrumental and transmission capacity of the early probes were only able to provide answers to some of the questions. Useful guidance on the nature of motion and heating of bodies entering a planetary atmosphere can be obtained from an approximate method published in 1958.

If a planet is considered exactly spherical, its radius (r) and surface gravity (g) define a natural satellite orbit speed, viz. $u_c = \sqrt{(gr)}$. Its atmosphere, to a first approximation, can be taken to be of uniform temperature (T) when, in the presence of the gravity field, the atmospheric density (ρ) falls off exponentially with height (h), viz. $\rho/\rho_0 = \rho^{-\beta h}$ with $\beta = Mg/R^*T$ where M is the mean molecular weight of the atmosphere and R^* is the universal gas constant. β can be calculated by observing the rate of decrease of brightness of a star as the planet's atmosphere passes across it.

Table 16

	Venus	Earth	Mars	Jupiter
Radius	0.97	1.0	0.53	11.0
Gravity	0.87	1.0	0.38	2.63
Relative deceleration	0.9	1.0	0.2	5.0
Typical re-entry temperature ratio	0.91	1.0	0.55	2.9
Relative heating rate	0.7	1.0	0.09	50
Relative total heat	0.8	1.0	0.2	50
Relative Reynolds number	1.0	1.0	0.4	2

From the equations of motion of a body entering an atmosphere (acted on by lift, drag, weight and centrifugal force), formulae are available for several quantities of importance in the design of spacecraft. The comparative heating and decelerations of ballistic vehicles entering some planets' atmospheres can also be evaluated (see table 16). These figures indicate that, from the point of view of severity of re-entry, Venus is very similar to the Earth, Mars considerably easier, and Jupiter much more difficult. However, it has been pointed out that quite different changes of pressure, temperature and speed of sound should be expected in the Martian atmosphere which is disturbed by wide temperature variations of its surface. Such considerations, and others yet unknown, may modify these comparative approximations.

Since this early approximation the actual measurement of planetary atmospheres has been transformed. A typical modern treatment of atmospheric entry is taken from the work of J. Jones *et al.* who proposed a means of calculating the radiative heat transfer to a vehicle entering the Venusian atmosphere. The lower atmosphere of Venus is now believed to consist of 97% of CO_2 with N_2 as the next most abundant gas. It was necessary therefore to determine which species of gas products would result behind the shock wave largely from CO_2 with also some lesser influences from CN, NO and N_2. To perform such a task requires a further detail level of behaviour of individual atoms which is dominated by their electrical properties since ions and electrons are evolved at the temperature encountered in passing through the shock wave. The definition of the atomic structure in such processes is far removed from all aerodynamic relations so far described and is more closely related to atomic physics. But, since the atomic behaviour changes as the gas passes through the flow field around the vehicle, different atomic and electronic energy transfers take place and both the theoretical treatment becomes most complex and the experimental confirmation of the many separate effects difficult and subject to uncertainties. The outcome of this analysis was a recommendation for twelve different relationships to allow for the most dominant effects.

Tektites

Tektites are small, natural, glassy objects of unknown origin which have been discovered at various parts of the Earth during the last 170 years. There has been much speculation about their formation, one theory suggesting, on the evidence of their chemical composition, that they are caused by impact of meteorites on soil. More recently an aerodynamic explanation has been offered showing how objects entering the Earth's atmosphere could be melted into stable shapes and thus take up the exact forms of tektites. The strength of this evidence appears to be overwhelming as artificial tektite glass spheres have been produced and placed in high temperature hypersonic air jets such as those used in the Mercury heat shield programme. The shapes of the artificial tektites are identical to the natural ones, even the waves and undulations of the melted front face and the thickness of the melted layer (as shown by microscopic examination of cross-sections). Another interesting experimental feature of this work was the simulation of ablation by use of frozen glycerin in a Mach 3 airstream. Still other heat transfer tests have been reported in which ice was melted in a subsonic airflow, the conditions being adjusted to give the correct modelling of heat transfer and melting of the surface.

This tektite analysis is a fascinating example of the work of aerodynamic research, and it is interesting to note that the melting point of tektite 'glass' is 2400 K which exceeds that of 'Pyrex' (melting at 2100 K) and ordinary soda glass, 1900 K. Nature has even anticipated man's attempts to discover materials capable of atmospheric re-entry!

Planetary atmospheres[38]

Since the first edition of this book was published in 1963 there has been a revolution in the study of the atmospheres of the planets. Not only have spacecraft encircled many of them and measured radiations from surface, clouds and the atmosphere itself, but probes have parachuted through the atmosphere and then acted as automatic weather stations for months on end. New kinds of radar on Earth have been able to supplement these data and elaborate planetary atmospheric mathematical models summate the information and produce total dynamic motions for comparison with observations. The volume of scientific reports and books on the subject is prodigious, running into tens of thousands, and could be seen as an explosion of interest similar to the new view of the Universe provided by Galileo's invention of the telescope in 1609. Some brief comparisons with features of the earth's atmosphere follow.

There is little scope for 'aerodynamics' on the planet Mercury. Its atmosphere of CO_2 and Argon has a surface pressure of only 1/1000th that of Earth and with a sunlit surface temperature of 700 K. Its magnetic field has been observed giving a small magnetosphere which is preceded by a bow shock as it passes through the solar wind.

The Venus atmosphere is a much richer brew with a surface temperature of about 740 K, a surface pressure of 85–100 Earth atmospheres and twenty constituent gases, mostly CO_2. Sufficiently detailed observations indicating zonal winds of the order of 220 mph (100 m/s) have been explained by physical processes not encountered on Earth. The atmosphere becomes strongly heated below the sun and a large convection cell is formed which rises obliquely and imparts lateral displacements sufficient to energise a large wind system. Similar effects have been shown in laboratory experiments with liquid mercury which exhibits a comparable Prandtl number. Clouds have been detected at many levels and curiously have been explained as halogen compounds of mercury which nevertheless do not appear as one of the twenty atmospheric constituents. Many alternative atmospherochemical models have been proposed for the Venusian atmosphere and refinement and argument must be expected for some time to come.

Mars also has an atmosphere largely of CO_2 (94%) but which is much thinner than that of the earth, i.e. 1/100th the density at the surface. Temperatures are lower, at the further distance from the solar radiation. In other respects it behaves much as does the Earth's atmosphere; there is a marked circulation of water, clouds of water vapour, ice and CO_2 circulate and there are daily and seasonal patterns of wind. Dust storms are frequent and, unlike those on the Earth, they occasionally grow in strength to envelop the whole planet. Distinctive cloud wakes downwind of volcanoes have been photographed. It is now recognised that the existence of dust and haze led to overestimation of the Martian atmospheric density quoted in the 1958 work.

The Jupiter atmosphere is strikingly different from those of the smaller inner planets. Its higher surface gravity enables it to hold substantial amounts of the light gases helium and hydrogen. Although there is general agreement on the basic internal structure of Jupiter there is still room for many alternative theories in detail. A central solid core is postulated of 7 500 miles (12 000 km) radius; then there is a vast depth of liquid metallic hydrogen under enormous pressure out to a radius of 28 000 miles (45 000 km). Above this lies another 15 500 miles (25 000 km) of liquid hydrogen above which is the atmosphere proper which is variously estimated to be from 60–3600 miles (95 to 5800 km) thick (most probable value is 620 miles (1000 km)). The lower surface is an indefinite transition zone where the pressure is millions of times that of the Earth's atmosphere. The atmosphere contains water droplets and ice crystals at the lower layers and ammonia (NH_3) crystals at intermediate levels with gaseous hydrogen at the top. It is characterised by very dynamic cloud motions which can be observed through the highest level (31–40 miles/50–65 km thick) which is transparent. The cloud patterns are very distinct, arranged in streams parallel to the equator; dark brown or grey stripes are termed belts with the yellow-white coloured bands between them called zones. Sixteen permanent belts and zones are named. The smaller detail of the clouds are at least as varied as those on Earth and contain streaks, wisps, plumes, spots and festoons (fig. 73). The belts rotate at

Fig. 73 The Red Spot of Jupiter, a very stable storm system (the smaller, lighter, spot is another, more mobile, storm)

different speeds and storm features characterised by white or brown spots which grow and decay rapidly. In contrast is the Great Red Spot first seen in 1664, which is variously 15 000 miles (24 000 km) to 30 000 miles (48 000 km) long, and is believed to be a very stable group of hurricane-like storms. This appears to be the highest cloud structure of Jupiter which may imply some additional internal energy. Already dynamic computer models have been prepared of the Jupiter atmosphere and starting from a generally static position of equilibrium quickly set up zones and belts which also reproduce many of the observed cloud motions photographed by passing spacecraft.

Jupiter has a strong magnetic field which traps solar wind charged particles in radiation belts like those named after Van Allen which surround the Earth.

Saturn's atmosphere was observed closely for the first time in November 1980 when the Voyager I spacecraft flew within 77 000 miles (124 050 km) of its cloud tops. Superficially similar to Jupiter its atmosphere has marked belts parallel to its equator but is believed to consist of methane (CH_4), NH_3, O_3 and sulphur dioxide (SO_2). Small scale convective clouds and wave like systems have been observed from spacecraft transmitted pictures and these may be of methane and ammonia. Voyager also passed within 2500 miles (4000 km) of Titan, the Saturn moon, which is the only moon having its own atmosphere. It found a thick haze layer 175 miles (280 km) thick with three separate haze layers above it. Voyager 2 will give another opportunity to study Saturn in 1981 after which both Voyagers head off towards Uranus about whose atmosphere little is known. It lies 1800 million miles (2900 million km) from the sun and is thought to have an atmosphere of methane and hydrogen.

Magnetohydrodynamics and plasmas

Magnetohydrodynamics (MHD), is the study of the motion of an electrically conducting fluid in the presence of a magnetic field. Examples of this type of flow have already been mentioned, e.g. in the rarefied upper atmosphere and the Earth's magnetic field. The practical interest in MHD for aeronautics and spacecraft arises from the ionisation of airflow past bodies moving at and above satellite speeds, and the prospect of producing MHD controlling forces, cooling systems and airflow measuring devices.

MHD effects are not only confined to the upper atmosphere near the Earth but are important in two other regions, viz. in the depths of space on the gigantic scale of stars and galaxies, and in very high temperature gas streams created on the Earth's surface in laboratories. In all these examples the conducting fluid is a gas at such a high temperature that the atoms are partially or totally stripped of their electrons. Matter in this condition is referred to as plasma, and represents the fourth state of matter–the other well-known states being the solid, liquid and gas forms. Although it is only in recent years that knowledge of the plasma phenomena and their occurrence has been unified, it is of growing significance as it is the state in which about 97% of the matter of the Universe exists and, in addition, it holds promise of unlimited world power production.

Magneto-aerodynamics

In aeronautics and spaceflight, MHD forces become comparable with aerodynamic forces when the magnetic and kinetic energy terms are of the same order of magnitude, i.e. $B^2/H\pi\rho\mu V^2 \approx 1$.† In atmospheric flight at $M = 20$,

† Here V = velocity, μ = magnetic permeability.

$h = 200\,000$ ft (61 km) the electrical conductivity of air (σ) is of the order of 30 mhos/ft (100 siemens/m) and the magnetic forces might then be about 1/10th of the kinetic. However, if small quantities of potassium (0.1%) are 'seeded' into the airflow, the conductivity could be increased ten times. In an MHD airflow there can be shock waves, vortices and a magnetic Reynolds number $Vl\mu\sigma$.

The application of MHD to aeronautics is still only in its infancy, but experiments have already shown some interesting effects. A magnetic field introduced into a cone in a shock tunnel flow displaces the shock wave and increases the drag. Some of this extra drag appears as Joule heat in the coil, but some is absorbed by energy changes in the air itself. Other experiments used solenoids placed across a hypersonic airstream to develop a lift force, as shown by distortion of the visible shock front and deflection of streamlines.

The use of magnetic fields in conjunction with electrically conducting boundary layers suggests a reduction in heat transfer to the surface as the magnetic field tends to suppress turbulence. Many theoretical studies have also been made of cylinders and aerofoils in conducting media. MHD waves can be set up by bodies moving fast at high altitude which distort the ionised wake and change the radar echo from it.

Plasma flow

While the significant MHD effects in aeronautics are confined to extreme speeds and altitudes, plasma flows at temperatures from 15 to 30 000 K are readily attained in the laboratory, and these are of such a high degree of ionisation that the electromagnetic effects become paramount. The first experiments were made in 1924 at a time when little practical use could be made of them, but the coming of rockets and spaceflight temperatures gave research a tremendous impetus. A great deal is now known about plasma flow, which can sometimes best be treated as a continuum process using averaged gas quantities, and sometimes by means of statistical mechanics. Extra properties are necessary to define the flow such as the magnetic field energy per unit volume $B^2/2\mu$, and the energy dissipation by Joule losses $(\nabla \times B)^2/\mu^2\sigma$. An analogous Mach number can be expressed as a ratio of field to thermal energies, viz. $M_b = B^2/2\mu p$, and the equations of motion can be represented in non-dimensional form.

There are at present four main applications of such plasma flows:

Plasma torch. Air, argon or helium is passed tangentially into a cylindrical chamber, across the axis of which is an electric arc. The swirling gas motion, combined with the high temperature of the ionised gas, produces a jet which is driven out by electromagnetic forces. The high temperature jet readily melts refractory substances.

Plasma jet. Such a jet of gas produces a reaction thrust which, although small (usually of only a fraction of a pound), is highly efficient for accelerating fast spacecraft. In early experiments, prior to 1960, ballistic pendulum techniques were used to measure the mass and momentum of microscopic bursts of plasma from electric discharges with a partly cut-away table-tennis ball as the mass. Experimental models of plasma propulsion systems launched into space during the 1960s. Many other possible means of converting electrical energy to jet thrust involve other MHD processes.

Plasma-driven wind-tunnel. Several electric arc-driven wind tunnels are now in use. Termed 'hotshots' or 'plasmadynes' they produce a very high temperature gas discharge by MHD effects and reproduce the high heat transfer properties of air at satellite and re-entry speeds.

MHD power generation. Electrical power has been generated by passing a high temperature (3500°C) combustion flame through a magnetic field and extracting electrical current from electrodes in the flow. The theoretical efficiency of a system producing electricity directly from a heat source depends on the temperature ratio in the process. The top value is limited by materials and the lower by the marked loss of conductivity below 2000 K. Other schemes arrange for the flow to pulsate, thereby giving a.c. power.

In thermonuclear plasma power generation, the plasma temperature must be raised to about 100M K, using heavy hydrogen at low pressure. This process is rather far removed from aerodynamics, but is of interest to the subjects described in this book as the gas flow is dominated by MHD effects and several of the conventional aerodynamic methods of flow description apply. Present interest centres on a means of keeping the hot plasma away from metallic containing surfaces (which would 'chill' the plasma) by magnetic fields. There are complex forms of instability in the plasma flow which have so far defeated attempts to make a successful magnetic 'bottle'. The target is attractive, however, for it has been estimated that thermonuclear power generation should yield an energy equivalent to 350 gallons (1.59 m^3) of petrol from the heavy hydrogen in one gallon of water. The heavy hydrogen occurs only as one part in 7000 parts of water. Lithium is also a candidate fuel for nuclear fusion power which is abundant in the Earth's crust and estimated to contain a fusion energy content of 10^{24} Joules (1.36 × 10^{24} ft lbf). This would last 5000 years at the 1978 world energy usage rate and is equivalent to an energy store six times as great as the total world fossil reserves. Annual expenditure on fusion research increased about ten times during the 1970s following the Mid-East oil crisis. Early steps taken in the 1950s to contain hot plasma in a toroidal chamber were unsuccessful (UK Zeta of 1958) but subsequent Russian work with TOKAMAK used a ring current in the plasma circuit itself which imparted the necessary stability. The plasma flow was so complex and difficult to measure that special laser plasma diagnostic techniques had to be developed during the 1960s before TOKAMAK was accepted as a breakthrough. During the 1970s several TOKAMAKs of increasing size and temperature were built culminating in the European collaborative venture JET (Joint European TORUS) installed in the UK[39]. The importance of JET in bridging between existing performance and that needed to sustain fusion reactions is indicated in fig. 74. The

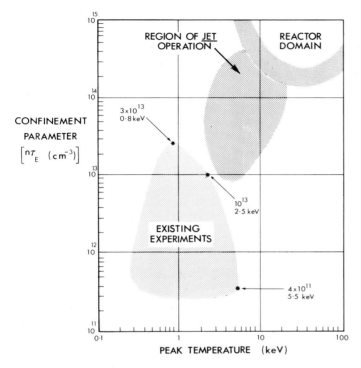

Fig. 74 Predicted range of plasma parameters for JET[39]. Three recent experimental results are indicated by asterisks

confinement parameter is the product of two quantities: η is the plasma density in ions/m^3 and τ_E is the plasma energy replacement time (seconds). (N.B. 1 keV = 1.16 × 10^7 K) Present evidence suggests that η should be limited to 3 × 10^{14} to limit neutron damage rate to the walls and maintain the peak plasma pressure to 10 atmospheres. Correspondingly τ_E is in the region of 1–2 seconds and has been shown to be difficult to attain in small sized experiments. JET should be fully operational during the 1980s but is only an experimental research machine. If it shows how stable fusion reaction plasmas can be produced there still remains the task of making a fusion power station that will collect the energy and feed it to the electrical supply grid. If only the heat can be used so that steam boilers and turbines are still required then much of the potential of the scheme will have been lost.

Quite other techniques of producing fusion power are under way employing laser, electron and particle beams. The essential idea is to focus incredible amounts of power on small fuel pellets containing heavy hydrogen and tritium to create an implosion with rapid rise of temperature and pressure similar to those attained in JET. The power beams are radially disposed to concentrate as much as 4 × 10^{10} hp (30 TW) on a pellet smaller than a salt grain. In this case also only crude heat is evolved still requiring some subsequent energy transfer process. The general availability of fusion power is not likely to emerge until well into the twenty-first century.

Cosmic electrodynamics

MHD effects are now known to play an important role in many cosmic phenomena and can, for example, produce electromagnetic forces exceeding those of gravity. Some of these phenomena are:

> Magnetic storms and aurorae in the ionosphere,
> ionospheric winds,
> solar flares,
> convective flow at the sun's surface (fig. 75),
> spiral arms of the galaxies,
> mixing of gas clouds in galactic collisions.

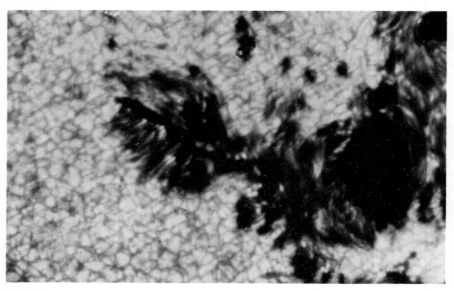

Fig. 75 Convection granulations and sunspots on solar surface

Some of the cosmic gaseous motions have flow patterns resembling those of terrestrial aerodynamics, e.g. turbulence, shock and gravity waves, vortices and convective granules.

The flow 'regimes' are determined by the physical conditions of the plasma which range over wide extremes (depending on temperature, density, magnetic field strength, etc.) and on the boundary conditions, which may be a cool, solid planet, in or around a star, or in the dusty clouds of the galaxies. The MHD flows in cosmic conditions are defined by four-dimensional partial differential equations; a task beyond present day solution. Modelling laws are deducible, however, which indicate that to reproduce similar flow parameters in a laboratory requires magnetic fields of thousands of gauss, whereas on the cosmic scale the field strength is very low, e.g. 20 gauss (2×10^{-3} T) in the solar corona and 10^{-6} to 10^{-12} gauss (10^{-10} to 10^{-16} T) in space itself. Very few MHD phenomena in space can be accurately represented in the laboratory, however, because of the difficulty of reproducing representative boundary conditions. This imposes restrictions on an experimental development of the subject, but the difficulty may be partly circumvented in time.

Alfvén and Fälthammer[40] compared laboratory and cosmic plasma characteristics in 1963, summarised in Table 17. The criterion for neglecting electromagnetic effects in gas dynamics is that the characteristic hydromagnetic parameter L is much less than unity, viz.

$$L = \frac{Bl\sigma_E P_B^{1/2}}{c^2 \rho^{1/2}} \ll 1$$

where P_B, σ_E, and ρ are the magnetic permeability, the electrical conductivity, and the density of the medium, respectively; B is the magnetic field strength: l is the linear extent of the medium; and c is the velocity of light. In cosmic problems involving interplanetary and interstellar phenomena, L is usually of the order 10^{15} to 10^{20}. In planetary ionospheres it reaches unity in the E layer.

The conclusion from these figures is that apart from the planetary atmospheres and oceans all fluid dynamic problems must account for electromagnetic effects. Three recent examples of such flows are given below.

Solar wind and planetary magnetospheres

The outward flow of particles and radiation from the sun is called the solar wind. A planet orbiting the sun passes across the solar wind stream. If it also possesses a magnetic field this creates a magnetosphere through which the solar wind cannot pass. A bow shock is created extending many tens of thousands of km and which can be measured by spacecraft's instruments. In the flow left behind the planet is a wake and there is a kind of turbulence at the boundary when this interacts with the solar wind.

Turbulence in interstellar clouds

The virtual emptiness of space is nevertheless populated by clouds of dust and gas of which some are luminous and others are not. Within our galaxy they account for 10% of its total mass and are important as they are the raw material from which new stars are created. An interesting type of cloud is termed a Bok globule, named after its Dutch discoverer of the 1940s.[41] They can be recognised as patches obscuring the background stellar luminosity and

Table 17 Characteristic quantities for laboratory and cosmic plasmas

	Linear dimension l (cm)	Magnetic field strength B (G)	Density ρ (g/cm^3)	Conductivity σ_E (esu)	Characteristic hydromagnetic parameter L
Laboratory experiments:					
Mercury	10	1×10^4	13.5	9.20×10^{15}	0.3
Sodium	10	1×10^4	0.93	9.37×10^{16}	10.8
Ionised gas (hydrogen)	10	1×10^3	1×10^{-10}	4.8×10^{14}	5.3×10^2
Cosmic plasmas:					
Earth's interior	2×10^8	10(?)	10	7×10^{15}	5×10^3
Sunspots	1×10^9	2×10^3	1×10^{-4}	4×10^{14}	9×10^7
Solar granulation	1×10^8	1×10^2	1×10^{-7}	7×10^{13}	2.5×10^6
Magnetic variable stars	1×10^{12}	1×10^4	1(?)	7×10^{15}	7.8×10^{10}
Interstellar space (more condensed regions)	1×10^{22}	$1 \times 10^{-5}(?)$	$1 \times 10^{-24}(?)$	$7 \times 10^{12}(?)$	7.8×10^{20}
Interplanetary space	1×10^{13}	1×10^{-4}	1×10^{-23}	$7 \times 10^{14}(?)$	2.5×10^{14}
Solar corona	1×10^{11}	1(?)	$1 \times 10^{-18}(?)$	$7 \times 10^{15}(?)$	7.8×10^{14}
Dark clouds	1×10^{13}	1×10^{-6}	1×10^{-20}	5×10^{12}	5.6×10^8

their density and motion can be inferred from the temperature of their constituent isotopic forms of CO_2 gas. Their mass, average temperature and size can be deduced and also the critical radius, which if they contract below this value, a gravitational 'condensation' will set in leading to the aggregation of a new star. Turbulence, or differences in velocity between circulating eddies in various parts of such clouds is a possible explanation of the mechanism that begins this creating process.

Shock waves in the interstellar medium

Yet another mechanism has been described by M. Zeilik. Taking the example of the Omega Nebula M 17 in Sagittarius he showed by means of carbon dioxide temperature-contours (5 to 50 K) evidence of a shock wave between two colliding molecular clouds of gas. In this region clusters of newly formed massive stars have been identified. The mechanism of the shock wave is to concentrate the gas locally and set up local gravitational instabilities. It is possible that both turbulence and shock wave effects in interstellar gases are at work, either separately or in combination.

Relativistic shock waves

In the study of the formation of novae, gases are found to have a propagation velocity measured as a fraction of the speed of light.

9
Aerodynamics and Civilisation

> Powers that will work for thee; air, earth and skies
> Wordsworth

The survey given in this book emphasises the fact that many present-day activities depend on aerodynamic knowledge. Whilst it could be argued that electricity and mechanical engineering were the big influences in the 19th century, aerodynamics belongs to the 20th century. It cannot claim to be the major influence but it shares, with electronic computers, nuclear energy, automation, radar and television, a leading place in modern science and technology. Aviation, both civil and military, has undoubtedly accelerated the pace of life but, like so many other developments, it is difficult, if not impossible, to weigh the good effects against the bad. Taking the long view, it may be that universal air travel will be seen to be one of the major advances in the breaking down of illogical barriers between nations. Without aerodynamic science this could not have come about; on the other hand, neither could the long-range multi-megaton nuclear rocket bomb.

Man-made changes to the atmosphere

There are many polluting gases and substances that now enter the atmosphere in increasingly large quantities. Burning coal, oil and gas in air gives off carbon dioxide and particulate matter which modifies temperature. How these quantities are assessed and how they diffuse into the air and alter its characteristics are of great aerodynamic interest. Although the mass of air is vast (approximately 1 million tons per inhabitant) both population and energy consumption are growing significantly. From UN statistics the world population is estimated to grow from a total of 3.7 billion in 1972 to 6.4 billion by the year 2000. The primary energy use for the non-communist world over the same period is expected to rise from 80 million barrels per day oil equivalent to between 160 and 207 (MBDOE).† The latter 25% range in predicted value made by the Workshop on Alternative Energy Strategies in 1977 is a typical uncertainty in forward projection of man's global activity.

† 1 MBDOE = 2.2×10^{18} J per annum.

Energy and heating effects[42]

The overall direct heating of the atmosphere by energy consumption is very small, viz. less than 0.01% of the total solar input. Industrial and city concentrations can however exceed the natural value. It has been calculated that a very large nuclear power station constructed over the ocean and liberating heat to the atmosphere at four times that of the solar input would create a large microclimate disturbance with increased rainfall extending 620 miles (1000 km) downwind. The most powerful influence comes via the CO_2 liberated in combustion by means of the so-called 'greenhouse-effect'. Carbon dioxide in the air absorbs the infra-red radiation emitted from the Earth's surface and hence the 0.5% per annum increase in CO_2 concentration due to burning fossil fuels is a matter for some concern. Doubling the CO_2 concentration from its present value of 334 parts per million (by volume) could raise the average atmospheric temperature as much as 3°C. Unfortunately the CO_2 story is far more complex and subject to great uncertainty at present. Although biomass absorbs CO_2 from the atmosphere the dominant effect in recent years arising from 0.5–1.5% per annum destruction of tropical rain forests has been to liberate as much CO_2 as does the burning of fossil fuels. Large quantities of CO_2 enter the soil and oceans and many of these global values are difficult to estimate even to an order of magnitude. A further difficulty is the part played by particulate matter, notably volcanic ash and aerosols. The effect of dust capable of absorbing 4% of the solar input was to locally increase temperature by 10°C in the stratosphere. In spite of these limitations some interesting non-linear models have been set up to investigate possible long term trends in atmospheric CO_2 and temperature arising from alternative global energy strategies. Niehaus reported[43] such a study in 1976 from the International Institute for Applied Systems Analysis in Austria. The object was to assess how different balances of coal and nuclear power used world-wide would influence atmospheric CO_2, etc. by the year 2100. The model took into account eight levels of activity including the atmospheric and oceanic circulation of CO_2, a biomass contribution, population and energy use predictions, industrial and resource activity and the dumping of the carbon radioactive isotope ^{14}C in the atmosphere as a result of nuclear power operations and processes.

Some typical results from this study are shown in Table 18:

This suggests that a non-nuclear strategy having to depend on a major use of coal for a substantial increase of world energy use would give an unacceptable atmospheric temperature rise. But this is an interim statement of its day which will be revised as more data are found on atmospheric circulations and the assumptions of the model are improved. It is nevertheless intriguing to recognise that scientists and engineers are making assessments of the possible future consequences of mankind's civilisation even out to 125 years beyond the present. The model showed that the biomass contribution was crucial and therefore should receive a high priority in future research. Marchetti (also of IIASA) pointed to an intriguing way of speeding up the

Table 18

Energy strategy	Optimistic nuclear fission-fusion	No nuclear power in future (mostly coal)	Coal with lightwater nuclear reactors	Reduced CO_2 strategy. Nuclear, coal out by 2100, Alt. sources
Maximum world energy use Φ 10^9 tCE/year	78	55	58	79
At date	2060	2050	2045	2070
A.D. 2100 CO_2 concentration (ppm*)	550	1400	1300	470
Temp. rise (°C)	2	9	7.5	1.4

In 1980: Φ11, *334. tCE = tons coal equivalent = 2.9×10^{10} J.

process of returning excess CO_2 to the deep Atlantic ocean by means of a natural concentrating mechanism which starts by increased salinity in the Mediterranean. A possible disadvantage of nuclear power, viz. an increase in radioactive ^{14}C was also estimated and could need as much attention for its future effect on the atmosphere as that of CO_2. It is not inconceivable too that with such good early warnings of these effects already published means could be found of burning coal in new engines which do not liberate CO_2 into the air at all.

Pollutants

In a sense CO_2 and volcanic dust could be termed pollutants although described in the previous section. Some pollutants have very serious local effects such as motor vehicle emissions in large cities and others are not yet understood well enough to be introduced into global atmospheric models. In this section a large number of such influences are touched on briefly.

Vehicle emissions. The concentrated poisonous exhaust fumes in the city of Tokyo are well known, typified by point duty policemen wearing gas masks. What is not generally known is APPS (Atmospheric Pollution Prediction Service) which used an aerodynamic flow computer to evaluate the pollution level, given the city building/road layout, the weather forecast and the expected (measured) traffic density. If severe levels were expected some hours warning would be given and various ameliorative measures taken. Vehicle emissions contain many substances some of which are potentially harmful, including smoke, hydrocarbon vapours, nitrogen oxides, carbon monoxide, sulphur dioxide and lead. Since the late 1960s increasingly restrictive legislation has been introduced to make vehicle manufacturers change engine designs to reduce pollutants. Some were crude add-ons which burnt up or catalytically altered heavily polluted exhaust gas. Others recognising that greater combustion efficiency led to cleaner exhausts came up with more elegant and simpler solutions offering better fuel consumption as well. There are special photochemical processes which can produce smog from exhaust gases, as in California, and here regulations are more severe than elsewhere: the toughest are in Japan. Whether improved engine efficiency and exhaust cleanliness will more than compensate for the still increasing road traffic volume is not clear. It is certain that arguments over the seriousness of the problem will continue for a long time. A current concern is the emission of particulate lead from motor car exhausts. Of course the total mass of pollutants is a major factor but so also is the aerodynamic dispersion mechanism and the constriction of city buildings in preventing natural dispersion. The increasing use of Liquid Petroleum Gas as a replacement for petrol in Japan is claimed to reduce pollution significantly.

The pioneering work of APPS has now been extended considerably and is generally termed Air Quality Mathematical Modelling, which takes account of all pollutant sources, both mobile and static. Melli has reported such a study applied to sulphur dioxide in the region of the city of Venice. Berkowicz and Prahm from Denmark have developed sophisticated theoretical relationships for the diffusion of gas in the turbulence, e.g. of the lower atmosphere. Atmospheric quality surveys are now regularly made from specially equipped aircraft and harmful emissions from one country can be traced as fallout in others over distances of thousands of km. Laser doppler radar methods (LIDAR) can measure the size and concentration of particulate matter at great heights. In parallel with this atmospheric modelling and concern of pollution hazards is an increase in the number of industrial plants that minimise the dumping of harmful pollutants into the air. In Sweden, for example, urban refuse recycling systems not only do not pollute the air, they use aerodynamic means to separate different constituents of the waste.

Pollution at high altitudes and the ozone layer. This is typified by two different effects, viz. the nitrogen oxides in the exhaust of the SST Concorde and certain halogen gases emitted at ground level which rise to stratospheric levels. All these chemicals interact destructively with the natural ozone (O_3) layer. This is created photochemically by incoming ultra-violet (UV) radiation and is generally to be found at altitudes where the atmospheric pressure lies between 10 and 100 mb (1 kPa and 10 kPa). Although only a minor constituent of the air, it has a profound meteorological significance such as the positions of the jet streams and on surface temperature. Since it strongly absorbs UV radiation it protects the biosphere from harmful radiation levels. Although its resilience to change and its capacity to deal with man-made chemicals cannot be known with certainty (over a hundred possible chemical reactions have been described), it is felt to be so important and possibly vulnerable that it is receiving considerable attention from many scientific authorities. Ozone density and distribution is measured from balloon sondes, by high flying aircraft (U2 and Concorde) and from satellites.

At the high combustion temperature in the jet engines of the Concorde (and other supersonic aircraft) nitrogen and oxygen of the air combine to form various oxides of nitrogen. Ozone combines with nitric oxide (NO) to form nitrogen dioxide (NO_2) and oxygen. The natural nitrogen cycle also creates nitric oxides from the biomass which also diffuse up to high altitudes. Combination with hydrogen produces nitric acid which is washed out of the troposphere by rain. Careful international research suggests that 500 Concordes each flying five hours per day would reduce the total ozone by no more than 1% which is well below the natural fluctuations. A more serious potential threat was revealed in 1976 when it was alleged that the increasing use of fertilisers for agriculture (which release nitrous oxide (N_2O) into the air) might eventually reduce the ozone layer by as much as 30%, seriously affecting the meteorological balance by the next century.

In 1974 Rowland and Molina suggested that Freon gases released from aerosol cans, refrigerators and air conditioning systems would reach the stratosphere to be decomposed by UV radiation liberating free chlorine which destroys the atmospheric ozone. Complex photochemical reactions are involved and estimates of the eventual likely effects vary over wide ranges.[44] Nevertheless the US Food and Drug Administration has seen fit to prohibit the use of these chlorofluorocarbons in aerosol canisters. Global weather models have been used to investigate the possible effect of ozone layer changes on air temperature and other climatic effects. A 50% reduction of ozone resulted in stratospheric cooling by 20°C but there were only trivial effects on temperature and rainfall at lower levels. The general consensus from all these investigations is that aircraft and freons are not likely to affect climate adversely during this century.

Spacecraft rocket exhausts dump significant amounts of chlorine compounds in the lower atmosphere around 5000 ft (1500 m) which are washed out by rain as hydrochloric acid (HCl). Interaction between gaseous and liquid HCl, rain droplets and other particulates of the exhaust (such as aluminium oxide) is a complex aerodynamic process which has been studied both in laboratory experiments and during actual rocket launches.

Aeronautics

Although the fundamentals of hypersonic aerodynamics have been well established and a possible future identified for very long range high speed aircraft, it seems more likely that preoccupations with increasing energy costs will be a dominant issue in the next two decades. Improvements of aerodynamic and propulsion efficiency will be two of the important contributions to hold the rising fuel costs in check. If the liquid hydrogen airliner appears, this is more likely after A.D. 2000 than before. Some of the highly desirable quiet intercity jet lift transports will also have to await the successful outcome of the efforts to maintain aeronautics profitably for the time being and are creations of the 21st century. Low speed agricultural aviation and air transport for the third world will also grow substantially.

The great step forward in man's achievement of flight was the separation of the means of lift from those of propulsion. It is interesting to speculate whether this trend could be reversed by combining the efflux of the propulsion system with the boundary layer of wing and fuselage in a form of wake propulsion. Although capable of significant theoretical gains in efficiency, attempts to achieve this so far have been frustrated by practical difficulties, not least of which is the intricate interaction between airflows which are traditionally dealt with separately by aircraft designers and engine manufacturers.

Noise reduction could also be enhanced by such an arrangement and this must continue to be a highly prized goal for future aircraft. Better understanding of the basic mechanisms of aerodynamically produced noise might lead to a favourable reduction of drag by suitably focussed noise sources but this is rather unlikely.

In the military field arguments still continue as to the virtue of flying faster than Mach 2. The driving force to improve performance and versatility is still strong against known and imagined threats and there is at present a plethora of new shapes under test. Jet lift aircraft capable of flying off from small areas and ships must be of continuing military value and some technical problems to improve them need better aerodynamics as well as new engines. Beam and particle weapons[45] using plasma and gas dynamic laser techniques appear to pose a somewhat vaguely understood threat to both aircraft and guided missile warheads. Much is speculative today but there is undoubtedly a great deal of effort being expended on this topic both in the US and the Soviet Union.

Since the whole of aeronautics depends on the air having just the right kind of low viscosity to induce vorticity for wing lift, it is to be hoped that deterioration of air quality in the distant future will not subtly erode this apparently magical gift of the gods!

Spaceflight

Although still a novelty at the writing of the first edition this is now a well established global activity. Resource, ocean and atmospheric monitoring of the Earth is a prime responsibility to assist mankind's development. Not least of the significant space investigations is a far better understanding of the physics of the sun and its plasma and MHD energy reactions. Perhaps new knowledge here will assist the quest to develop nuclear fusion power on Earth. This is probably one of mankind's greatest challenges today for if it could not only provide plenteous and cheap power in the future, but also avoid burning and polluting the atmosphere, future generations may be grateful that atmospheric quality had been maintained.

After the Space Shuttle there is a prospect for an even larger kind of re-usable aerospace launcher which would be needed for another type of future energy system–the SSPS (Satellite Solar Power System). This has been extensively studied by the Department of Energy and NASA in the USA and the concept proposes the construction of large solar energy collecting arrays in orbit whose power is transmitted to receiving stations on Earth by microwave radio power. Perhaps 20% of the world's energy needs could be provided by such a system. Such is the present state of aerodynamic knowledge that already significant conclusions have been drawn of the likely effect of such a system on the atmospheric quality. The large rocket launch vehicle will dump large quantities of water vapour and nitric oxides in the lower atmosphere which is expected to alter cloudiness and precipitation. Stratospheric CO_2 and water injection is not anticipated to have any significantly deleterious influence on either the greenhouse effect or on the ozone layer. At higher altitude a 15% modification to the water concentration profile is admitted and the rocket exhaust plasma is expected to create large holes in the ionospheric F-regions which could affect high frequency radio propagation. The heating and electrical effects of the microwave beams are difficult to predict in view of the complex interactions in the upper levels of the ionosphere. The effect on human life on the surface is generally believed to be acceptable. It would be premature to make definite pronouncements of these polluting effects with present day knowledge but it is noteworthy that such environmental impacts are now fully taken into account in parallel with the other features, e.g. vehicle design, power production potential, electrical system design, long term programmes and overall costs.

Meteorology

A great deal is still to be done in improving the computer models of the atmosphere and deriving better short and long term predictive methods. As man moves into hitherto unpopulated regions various modifications of local climate may be expected. Improved irrigation and cultivation of vegetative zones on the borders of arid regions could lead to gradual long term changes in climate. There is a trend towards more human activity offshore following recent successful developments of sea based oil and gas fields. Energy systems such as wave power, nuclear power stations, wind generators and eventually perhaps SSPS receiving antennae (30 km diameter) could modify the microclimate. Further additions of power using and process industries and living accommodation for their operators would accentuate this trend.

The subject of weather modification and control, although it has attracted a lot of research and even more publicity for a long time, seems no nearer to solution. The energy content of the atmosphere is enormous and the cost of brute force additions of energy prohibitive. The good and bad effects of weather at any particular locality are hard enough to define, let alone change, and the political conflicts that would arise if one nation believed another had improved its own weather at the expense of the other are best avoided. A more expedient way of dealing with the worst effect of droughts and hurricanes is better shorter and longer term forecasting and a new kind of massive world relief agency that could attend to the problems in good time.

Unfortunately new weapons still plague mankind and in predicting the likely lethality of chemical and nerve gases meteorological knowledge of wind distribution and diffusivity in the boundary layers of the atmosphere is used.

The explosion of the Mount St. Helens volcano in the USA in May 1980 placed 200 000 tons (203 Gg) of volcanic material into the stratosphere, an increase of a quarter of the total already there. The scientific community was well prepared for the event and exploration of the plumes of gas and volcanic dust were made from aircraft, balloons, satellites and ground LIDAR in many countries. The effects on the ozone layer, stratospheric temperature and the ejection of unexpected gases all have meteorological implications. No one welcomes such a natural disaster, but, it having happened, it has become the most well documented volcano in history and could provide invaluable evidence of the response of the atmosphere to pollutants.

Aerodynamic techniques

In spite of the vast range of experimental equipment now in use which can represent virtually any airflow regime, apparatus is continually being improved. The range of validity of any one wind-tunnel is usually quite narrow so that several different models are necessary, especially for craft with a wide speed range, and accuracy of flow measurement could be bettered, e.g. at hypersonic speeds. The time taken to get experimental results is often inconveniently long, not because the experiment itself (usually fully automated) takes much time, but because it is a slow job to build the models.

A new class of low speed pressurised tunnel has been developed to give correct flight Reynolds number of about 6 million. The test section diameter is typically 16.4 ft (5 m) so the models are large and can reproduce very accurately fine details of the wing such as flaps and slats. A surprising result is a marked loss of maximum lift coefficient in the landing configuration and speed due to Mach number effects. Although the main flow speed is only of the order of M 0.2 flow over the high incidence leading edge slat closely approaches sonic speed. Proposals have been drawn up for large wind tunnels powered by water pressure held in large reservoir tanks and driving air above pistons through the tunnel section. In a Ludwieg tube tunnel a large cylindrical tube stores compressed air which creates a high speed flow past a model when exhaust valves are rapidly opened at the other end behind the model. ONERA in France has developed a beautiful technique of flow visualisation in a moving water tunnel by employing coloured streams of condensed milk which flow around a test model. This has been used for aircraft, helicopters, trains, cars and buildings. Since many colours can be used and the milk streams do not disperse as rapidly as does smoke in air the intricate interplay between boundary layer and vortices can be dramatically displayed. It would be a most encouraging prospect for the aerodynamic pioneers of the last century to be able to appreciate the impressive (and expensive) experimental facilities now available all over the world!

In the first edition the potential growth of aerodynamic computing power was recognised and it was speculated whether developments in this field would ever replace wind tunnels which are admittedly inexact both for aerodynamic and other practical reasons. The concept of an Ultimate Aerodynamic Computer was proposed in reference 46 on the basis of a computer which reproduced the behaviour equivalent to that of an actual air molecule. However, since the air molecule can respond to changes of movement and energy in twenty-eight different modes, and bearing in mind fundamental limitations in solving even Navier Stokes equations this now seems unlikely. Instead a new kind of relationship has been established between computer and wind tunnel so that, whereas in 1960 there were organisations exclusively devoted to experimental aerodynamics, most of these now include also an aerodynamic computer facility as well. In many such organisations the volume of computing work can be as great or even somewhat greater than the experimental effort.

Stop press

The rate of progress in aerodynamic matters is now so high that it is difficult to complete a book such as this because of late arrivals on new topics! To overcome this difficulty this section includes brief mention of some of them.

Ink-jet printing employs minute drops of ink which are squirted from a nozzle towards the paper and steered in flight at the command of a computer to create characters.[47] One hundred thousand drops are created each second and each one receives its own electric charge which steers it across a lateral magnetic field. By this method printing rates of 1000 characters per second are achieved. The aerodynamic drag of the minute drops has to be allowed for in the design of the printer. Unwanted aerodynamic disturbances cause migration of dots and loss of character quality.

An even more surprising application of this technique is a proposal to employ it to detect the presence of the quark–that elusive constituent of the elementary particles of matter.

Freeze drying of vegetables is now an essential part of civilisation. In order to freeze peas they are carried along a channel over which is blown an icy blast of air. It was observed that occasionally a pea would rise from the channel as aerodynamic lift was created over its upper curved surface. Once airborne the pea cooled very rapidly as it was completely bathed in the cooling airflow. Thus the peas cooled more rapidly, and individually, so they were more readily cooled than if they were in a massive block. Moderate redesign of the process turned an unwanted fault into a positive advantage by adjustment of the air velocity of the cooling blast!

During the 1970s a new means of aerodynamically separating gaseous isotopes emerged called the Jet Membrane Concept. An objective was a less costly means of separating uranium isotopes for nuclear power purposes. It employs a principle called background penetration separation which operates in rarefied mixed gases. A jet of the gas mixture is expanded through a nozzle into a chamber and an open ended tube centred along the jet centre-line collects more energetic particles that are able to travel upstream against the main jet flow. Although this seems a rather hit or miss process it has been shown both theoretically and experimentally to be effective.

Aerodynamics in nature is a continuing subject of research. The *Scientific American*[48] of December 1980 describes how the cockroach senses the approach of a predator by an aerodynamic process. An enemy approaching the potential victim moves towards it and in so doing displaces a volume of air. The minute wind so created is detected by many sensitive hairs hanging vertically from two spikes mounted at the cockroach's rear end. Activation of these sensors causes the cockroach to turn and run quickly in the direction away from the moving object. Wind speeds as low as 1.6 in/s (4 cm/s) are readily detectable although a toad's tongue flicking out has been measured to create a wind of 8 in/s (20 cm/s). The cockroaches seem able to distinguish the acceleration of the air. A value of 24 in/s^2 (60 cm/s^2) is a signal for rapid escape.

This last story reminds me of a theory I have held for some time concerning the alleged benefits to plant welfare resulting from talking to them. I tend not to discount stories which I cannot accept on a first hearing and I suspect that the real cause of this phenomenon is essentially aerodynamic. In going up to a plant to talk to it air is displaced, CO_2 and water vapour are breathed around it which I suspect improves its microclimate. Often people who talk to plants keep them in little-used rooms and it is possible that the plants suffer from lack of air circulation.

Conclusion

In this book, several aspects of aerodynamic science have been described and examples of its diverse applications given. The variety of motions of the air seems limitless, and new natural laws are still being discovered surprisingly often. The following thought is offered as an exercise for the reader on the many different frames of reference of aerodynamics. Consider a man flying in an aeroplane. He is breathing air. His skin is perspiring and transferring heat to the air. The cabin air-conditioning system is changing the outside air to suit his requirements. The aeroplane is forcing its way through the atmosphere in a jet stream which is itself moving at 200 mph within a much larger air mass, hundreds of miles across, which is travelling over the oceans. The Earth lies within the Sun's atmosphere which, in turn, is like an island in the infinite ocean of the universal plasma. Each of these eight frames of reference defines a very different kind of aerodynamics, and all are occurring at the same time, yet not interfering very much with each other. How many far larger or far smaller frames of reference have we still to find?

For some, the interest in aerodynamics is to discover the law and order of nature by theory and experiment, while for others it may be the desire to master some aerodynamic process. But for all of us there is a fascination in observing those aerodynamic motions common to our every-day experience, and to the behaviour of storms, sun and stars. Such universal laws give us an insight into the wonder of creation.

References

1. Von Kármán, Th. (1954) *Aerodynamics*. New York: Cornell University Press.
2. Glauert, H. (1947) *The Elements of Aerofoil and Airscrew Theory*. Cambridge: Cambridge University Press.
3. Richardson, E. G. (ed.) (1960) *Aerodynamic Capture of Particles*. London: Pergamon Press.
4. Goldstein, M. E. (1974) *Aeroacoustics*. USA: NASA SP 346.
5. Lighthill, M. J. (1952) 'On sound generated aerodynamically.' *Proc. Roy. Soc.* **A211,** 564–87 and (1954) **A222,** 1–32.
6. Krasnov, N. F. (1971) *Aerodynamics*. Translated for NASA (1978). TT 74-52006. NASA TT F-765. Moscow.
7. Clarke, J. F. (1978) 'Gas dynamics with relaxation effects'. *Rep. Prog. Physics* **41.** London: The Institute of Physics.
8. Krause, E. (1974) 'Application of numeral techniques in fluid mechanics'. *The Aeronautical Journal R.Ae.S.* London, August.
9. Welch, J. E., Harlow, F. H., Shannon, J. P. and Daly, B. R. (1965) 'The MAC method'. *Los Alamos Scientific Laboratory of the University of California.* Report No. LA-3425 (Revised). November.
10. Trolinger, J. D. (1974) 'Laser instrumentation for flow field diagnostics'. *AGARD–AG–186.* Neuilly sur Seine, France (March).
11. Scorer, R. S. (1978) *Environmental Aerodynamics*. Chichester: Ellis Horwood.
12. Brown, R. A. (ed.) (1976) 'The Union City, Oklahoma, tornado of 24 May 1973'. *NOAA Technical Memorandum ERL NSSL-80.* National Severe Storms Laboratory, Norman, Oklahoma, USA.
13. Costen, R. C. (1970) 'An equation for vortex motion including effects of buoyancy and sources with applications to tornadoes'. *NASA TN D-5964.* USA, Washington.
14. Mason, B. J. (1971) *The Global Atmospheric Research Programme–a Contribution to the Numerical Simulation and Prediction of the Global Atmosphere.* Earth-Science Reviews 7. Amsterdam: Elsevier.
15. Hutchinson, E. G. (1970) 'The biosphere'. *Scientific American* **223** (3), 44.
16. Simiu, E. and Scanlan, R. H. (1978) *Wind Effects on Structures*. New York: John Wiley.
17. Vogel, S. (1978) 'Organisms that capture currents'. *Scientific American* **239** (2), 108.

18. Bahadori, M. N. (1978) 'Passive cooling systems in Iranian architecture'. *Scientific American* **238**(2), 144.
19. Hoerner, S. F. (1958) *Fluid Dynamic Drag* (2nd edn). New Jersey: Hoerner.
20. Stollery, J. L. and Garry, K. P. (1980) 'Keep on trucking–but please change the shape'. College of Aeronautics, Cranfield, UK.
21. Gawthorpe, R. G. (1978) 'Aerodynamics of trains in the open air'. *Railway Engineer International*, London: Institution of Mechanical Engineers.
22. Wheeler, R. L. (1976) 'An appraisal of present and future large commercial hovercraft'. *The Aeronautical Journal, RAeSoc.* (August).
23. Chesters, J. H. (1953) 'The growth of an idea'. *Iron and Coal Trades Review* (23 January).
24. Humphrey, E. F. and Taramoto, D. H. (eds) (1965) *Fluidics*. Boston, Mass: Fluid Amplifier Association Inc.
25. Lighthill, Sir. J. (1974) 'Aerodynamic aspects of animal flight'. *5th Fluid Science Lecture*, London: The Royal Institution (20 June).
26. Weis-Fogh, T. and Jensen, M. (1956) 'Biology and physics of locust flight'. *Proceedings of the Royal Society of London, B,* **239**, 415–585.
27. Weis-Fogh, T. (1975) 'Unusual mechanisms for the generation of lift in flying animals'. *Scientific American* **233**(5), 81.
28. Hess, F. (1968) 'The aerodynamics of boomerangs'. *Scientific American* **219**(5), 124.
29. Mehta, R. and Wood, D. (1980) 'Aerodynamics of the cricket ball'. *New Scientist*, 7 August.
30. Vittek, J. F. Jr (ed.) (1975) 'Proceedings of the Interagency Workshop on lighter than air vehicles'. M.I.T. Flight Transportation Laboratory Report R75-2, USA.
31. Hoerner, S. F. and Borst, H. V. (1975) *Fluid-dynamic lift*. Hoerner Fluid Dynamics, P.O. Box 342, Brick Town, NJ 08723, USA.
32. Küchemann, D. (1978) *The Aerodynamic Design of Aircraft*. Oxford: Pergamon.
33. Barrère, M. L. (ed.) (1975) 'Analytical and numerical methods for investigation of flow fields with chemical reactions, especially related to combustion'. AGARD Conference Proceedings No. 164. Neuilly sur Seine, France.
34. Yaggy, P. F. (ed.) (1973) 'Helicopter aerodynamics and dynamics'. *AGARD Lecture Series No. 63*, Neuilly sur Seine, France.
35. The Hawker-Siddeley Harrier. Reprint from *Aircraft Engineering*, UK. Dec. 1969–April 1970.
36. Morgan, M. B. (1972) 'A new shape in the sky'. *The Aeronautical Journal*, London: The Royal Aeronautical Society.
37. Allen, H. J. (1967) 'Some problems of planetary atmosphere entry'. *The Aeronautical Journal*. London: The Royal Aeronautical Society.
38. West, G. S. Jr, Wright, J. J. and Euler, H. C. (eds) (1977) 'Space and planetary environment criteria guidelines for use in space vehicle development'. *NASA Technical Memorandum 78119* Alabama, USA: George C. Marshall Space Flight Centre.
39. Gibson, A. (1978) 'The JET project, a step towards the production of power by nuclear fusion'. 'In *Energy and Aerospace Conference*, London: The Royal Aeronautical Society.

40. Alfvén, H. and Arrhenius, G. (1976) *Evolution of the solar system*. SP-345, Washington DC: NASA.
41. Dickman, R. L. (1977) 'Bok globules'. *Scientific American* **233**(6), 66.
42. Mason, B. J. (1977) 'Man's influence on weather and climate'. *J. Royal Soc. Arts* (Feb), 150.
43. Niehaus, F. (1976) 'A non-linear eight level tandem model to calculate the future CO_2 and C-14 burden to the atmosphere. RM-76-35, Laxenburg, Austria (May): *International Institute of Applied Systems Analysis*.
44. Sugden, T. M. and West, T. F. (eds) (1980) Chorofluorocarbons in the environment: the aerosol controversy. (Society of Chemical Industry, London). Chichester: Ellis Horwood.
45. Parmentola, J. and Tsipis, K. (1979) 'Particle-beam weapons.' *Scientific American* **240** (4), 38.
46. Allen, J. E. (1971), 'The future of aeronautics–dreams and realities'. *The Aeronautical Journal, RAeSoc*. London (December), p. 587.
47. Kuhn, L. and Myers, R. A. (1979) 'Ink-jet printing'. *Scientific American* **240** (4), 120.
48. Camhi, J. M. (1980) 'The escape system of the cockroach.' *Scientific American* (December), **243**(6) 144.

Index

Entries in **bold** indicate main reference

Ablation, 166, 168, 169, 175
Acoustics, 23
Aerials, 81, 82
Aeroacoustics, 24
Aerodynamics, categories of (*see also* Flow), 3, 19, 35–7
   industrial, 102–17
   natural, 50–86
   of aeroplanes, 129–38, 145–60
   of animals, 118–23
   of propulsion, 138–41
   space, 161–85
   terrain, 67–8, 70
   transport, 87–102
Aerodynamics, history of, 4–7
Aerofoil theory, 6, 7
Aerogenerators, 112–17
Aeronautics, 2, 6, 118–60, 190–1
Aerothermochemistry, **40–2,** 168, 170, 172, 174–5
Aerothermoelasticity, 92
Air conditioning, 88, 91, 105, 130, 165, 195
   in nature, 83–6
Aircraft designs, 130–1
Airships, 126–9
Ampère's Law, 44
Andrean wind Turbine, 114–15
Animal flight, 118–23
Apollo, 168
APPS, 188–9
APT (train), 92–3
APT (satellite transmission), 62
Area rule, 39, 150–1
Aristotle, 4

Astrophysics, 2, 173
Atmosphere (*see also* Meteorology), 1, 161, 163, 173–5, 186–93

B.Ae., 145–6, 152
Ballistic pendulum, 5, 36, 179
Ballistics, 118, 162, 165–7
Bats, 120
Beam and particle weapons, 191
Bénard cells, 52, 182
Bernoulli's theorem, 11, 37
B.H.C., 98–101
Bicycles, 123
Biosphere, 65–7, 189
Birds, 119–20
Bladud, 4
Boomerangs, 125
Boost-gliders, 158, 169–72
Boundary layer, 7, 10, **14–15,** 50, 68, 102, 150, 154, 165, 191
   control, 136, 149, 155
   laminar, 14, 15, 136, 158, 166
   turbulent, 15, 136
Bridges, 76, 80, 81
Buildings, 75–80, 85–6
Buoyancy, 22, 37, 57, 103, 128

Calculations, *see* Computing
Carbon cycle, 66, 187–9
Carbon dioxide, 66, 67, 83–5, 187–9
Channel tunnel, 95
Chimney stacks, 80–1, 83
Chlorofluorocarbons, 95
Circulation (*see also* Flow, Meteorology, Vortex), **13,** 51–2, 131–3

INDEX

Climate, 58–67, 69–70, 195
Cloud box, 57
Clouds, 52–3, 57, 60, 62, 63, 64, 176–8
Cloud street, 52, 63
Coanda, Henri, 110
Cockroach, 195
Compressibility, **31**, 37–40
Computing, 2, 13, 27, 42, **45–8**, 58–65, 72, 95, 100, 125, 168, 179, 187–9, 194
Concorde, 152–7, 189
Conformal transformation, 13
Contrails, 137
Convection, **21**, 52, 57, 104, 165
Cooling (*see also* Heat transfer), 158, 165
Cosmic electrodynamics, 182–5
Cranfield Institute of Technology, 90
Cricket ball, 125–6
Crop spraying, 72, 107

Damköhler number, 41–2
Decibel, 24
Density, 1, **30**, 39, 163–5
Diffusion coefficient, 31, 71
Dimensional analysis, 33–5, 172
Distortion, 10
Doppler, 49, 53–4, 65, 189
Drag, 14, 16, **20**, 21, 39, 73, 82, 88–91, 93, 102, 120, 123, 124, 129, 132, 136, 138, 143, 151, 161–2, 164–5
Dusty gases, 42

Eddies, 10, 14, 77, 185
Ekman spiral, 75
Electrical analogy, 11–12
Electrostatic drag, 164
Emissivity, **32**, 166
Energy-natural, 51, 84, 186–8
  -sources: wind, 112–17
    fusion, 180–1
    wood, 72
    hydrogen, 157, 190
    SSPS, 191–2
  -transport, 90, 92, 97, 100
Environmental Power Assistance, 97
Equipotential surface, 11
ESSA, 62
Evaporation, 69, 71–2
Explosions, 19, 23, 51

Fans, 98, 140
F.A.R., Part 36, 28

Faraday's induction law, 45
Ferrybridge, 78
Fick's Law, 32
Fire, 83
Flight corridor, 160
Flow, 8–10
  continuum, 11
  free molecule, 32, 37, **44**
  hypersonic, 37, **42–4**, 157–60, 169–72
  Newtonian, 37, **44**
  patterns, 10–19, 182
  subsonic, **37–8**, 132–7
  supersonic, 14, 37, **39–40**, 150–7
  streamlined, 9–10
  transonic, 37, 39
  turbulent, 14, 68
  visualisation, 8–9
Fluid dynamics, 2, 37
Fluid elements, 10
Fluidics, 108–12
Flutter, 80, 136
Flying bedstead, 145, 150
Fog, 83
Forces (*see also* Drag, Heat, Lift, Pressure), 19–22, 33–4
Forest aerodynamics, 72–5
Fourier's law, 31
Frisbee, 125
Fronde efficiency, 138
Fronde number, 34
Fujiyama, 68
Furnaces, 102–4

Gale damage, 73
Galloping, 80
GARP, 58, 62
Gas constant, 17
Gas dynamics, 7
Gibraltar, 68–9, 78
Gliders, 67, 158
GOES, 62
Gossamer Albatross, 124
Gossamer Condor, 124
Grashof number, 34
Gravity, 30, 50, 174
Greeks, 4
Guided missiles, 138, 157, 161–2, 165

Hail, 57
Harrier, 145–50

INDEX

Head, total (H), 11
Heat energy (*see also* Energy), 19–21, 138–40
Heat Shields, 166–70, 173
Heat transfer, **21,** 130, 139–41, 154–5, 157–8, 159–60, 165–6, 168–70, 173–5, 178–82
Helicopter, 142–5
Helistat, 128
Helium, 127, 161, 176
Hen feathers, 8
Holograms, 49, 171
Hovercraft, 98–102
Hurricanes, 52, 193
Hush-kits, 27
Hydraulic analogy, 8
Hypersonic (*see* Flow)

Icarus, 4, 124
Ice, 137, 156, 176–7
I.I.A.S.A., 96, 187
ILLIAC IV, 46
Inertia, 19, 33
Inkjet printing, 194
Insects, 120–3
Inflatables, 136–7, 172
Interference, 13, 39, 40, 150
Ionisation, **32,** 43, 162, 165, 168, 178
Ionosphere, 161, 163, 192

JET, 180–1
Jet engines (*see also* turbojet), 138–41
Jet lift, 145–50
Jet streams, 51, 189
Jodrell Bank, 82
Joukowski, 5, 6, 132
JPL, 18

Kármán, von, 110
  vortex street, 16
Kinematic viscosity, 31
Knudsen number, 34
Küchemann, Dietrich, 134, 153

Laplace's equation, 11–13
Lasers, 48, 49, 65
Launching, 162, 164, 169, 171–2
Lewis number, 34
Leonardo da Vinci, 4, 124, 142

Lift force, 5, 6, 12, 13, **20–1,** 82, 118–19, 125, 127, 129, 132–3, 142–3, 145, 151–5, 158–60, 164, 168
Lighter than Air (LTA) (*see* Airships)
Lighthill, Sir James, 27, 118
Liquid hydrogen, 157, 190
Locusts, 69, 120–1
Lorentz force, 45, 164
Lotus, 88, 92
Ludwieg tube, 193

MAC, 47–8
Mach number (M), **34,** 37–8, 131, 134, 150–2, 154, 157–60, 171, 173, 178–9
Magnetic fields, 9, 11, 44–5, 164, 178–80, 182–4
Magnetic permeability, 32
Magnetohydrodynamics (MHD), 19, 22, 32, 37, **44–5,** 161, 168, 178–85
Magneto-aerodynamics, 178–9
Magnus force, 14, 97
Man-powered flight, 123–4
Marchetti, Cesare, 96, 187
MAVIS, 46–7
Mechanics of flight, 21
Mercury capsule, 167–8
Metals, 154–5, 157–8, 166
Meteorology, 50–70, 192–3
Microclimate, 195
Micrometeorology, 38, 69–75
Military aeronautics, 6, 161, 193
MIT, 6
Modelling, 35–45, 162, 183
Molecules, 10, 32, 44, 163
Monte Carlo, 47–8
Mount St Helens, 193
Motor cars, 87–92

Navier-Stokes equation, 45, 47, 194
Newton, Sir Isaac, 4–6, 31
Newtonian flow, 37, **44**
NIMBUS, 61, 64, 65
NOAA, 53–4, 62
Noise (*see also* Aeroacoustics), 23–9
  footprint, 28
  of Concorde, 152, 156
  of helicopter, 144
  of space shuttle, 171–2
NSSL, 53

Nuclear energy, 51, 180–2, 187
Nusselb number, 34

Ohm's law, 45
Orbits, 61–2, 164
Organ pipes, 16
Ornithopters, 4, 124
Oxygen, 1
  cycle, 66–7
Ozone, 156, 172, 189, 190

Paragliders, 160, 172–3
Particles, 22, 107
Passive cooling, 85–6
Péclet number, 34
Pegasus engine, 146, 148
Planets, 173–5, 182
Planetary atmospheres, 175–8
Plasma (*see also* MHD), 179–82
Plastics, 166
Pneumatic car, 92
Pollution, 61, 67, 70, 72, 83, 186–90, 192–3
Potential function, 11
Prairie dog, 83–4
Prandtl, Ludwig, 15
Prandtl-Glauert rule, 38
Prandtl number, 34
Pressure, 8, 11, 16–21, **30**, 130, 136, 156, 158
Pressure coefficient, 20
Propeller, 138–9
Propulsion, 129, 138–41, 159–60, 191

Quark, 194

Radiation, 50, 63–5, 164, 166, 174
Radioisotope, 173
Rain, 37, 53–4, 91, 172
Ramjets, 141, 159
Rayleigh, 5
Re-entry, 165–75
Relative density, 30
Relaxation time, 41–2
Reynold's number, **34–5,** 39, 47, 71–2, 77, 80–1, 84, 88, 102–3, 111, 121–2, 127
Robins, Benjamin, 5, 112
Rockets, 80–1, 141, 162, 169–72
Rolls Royce, 145–6

Rotation, 10
Rotorcraft, 141–5
Roughness, 27, 32, 35

Safety factors, 131
Satellites, **16–17,** 164–5, 191, 193
  meteorological, 61–5
  navigation, 97
Schlieren, 49, 173
Schmidt number, 34
Ships, 96–7
Shock waves, 16–18, 23–4, 39–40, 149–59, 165–7, 170, 183, 185
Skin friction, 14, 15, 35, 136
Smog, 69, 189
Smoke, 8, 22, 83, 96
Snow, 70–1, 137
Soil erosion, 70, 147
Sonic bang, 24, 152, 156
Sound power level, 24
Sound pressure level, 24
Sound, speed of, 16–18
Sources and sinks, 13
Spaceflight, 1, 161–83, 191–2
Space Shuttle, 169–72
Specific heat, 31
Sprays, 72, 105–7
SSPS (Satellite Solar Power System), 191–2
Stability and control, 44, **129–30,** 145, 150, 168, 170
Stall, 132–3
Stanton number, 34
STOL (Short Take-Off and Landing), 147
Streamlines, **9,** 10, 13, 68, 76–7, 79, 89, 106, 148–9
Subsonic (*see* Flow)
Sulphur, 52, 66
Sun, 1, 52, 164, 182, 191
Supersonic (*see* Flow)
  combustion, 159
Swept back wing, 134–6, 144
Swing wing, 151–2, 155

Tektites, 175
Temperature (*see also* Heat), 17, **30,** 39–43, 83, 139–40, 154–8, 165, 168–9, 172, 180
Termites, 84–5

INDEX

Theoretical methods, 1, 2, 4–7, 32–6, 45–9
Thermal conductivity, 31
Thermal diffusivity, 32
Tilt wing, 144
TIROS, 61, 62
TOKAMAK, 180
Tornadoes, 53–7
Trains, 92–6
Transition, 15
Translation, 10
Transonic (*see* Flow)
Transpiration, 69, 158
Turbojet, 24–9, 139, 140, 141, 145–50, 156–7
Turbulence, 4, 25, 68–9, 183–4
Turbulence amplifier, 110–11

UAC (Ultimate Aerodynamic Computer), 194
Ultrasonics, 23

V/STOL (Vertical and Short Take-off and Landing), 145–50
VTOL (Vertical Take-Off and Landing), 141, 144–5, 149
Varibale geometry, 151–2
Velocity, 8, 14
  potential, 11
Venusian entry, 173–5
Viscosity, 14, 15
Visualisation, 8, 18, 69, 153, 167

Vortec Corporation, 106
Vortex, **14,** 16, 52–7, 77, 80–2, 103, 132–3, 154–5
  Street, 16, 47
  Tube, 105–6
Vorticity, 14, 16

Wakes, 10, 16, 96, 141, 165–7
Water cycle, 66–7
  flow, 8, 53, 69, 103
Weather equations, 58–9
  forecasting, 58–65
  records, 60–1
Whirling arms, 5, 38, 112–13
Wind, 50–61, 67–86
  driven circulations, 72
  energy conversion systems (WECS), 112–17
  induced discomfort, 78–80
Windmills, 112–13, 117
Wind tunnels, 27, 36, 38–40, 44, 57, 68, 69, 87–93, 96, 120–1, 123–4, 137, 144–5, 154–5, 168, 171, 173, 175, 180, 193–4
Wing shapes, 119–23, 130–6, 151–5, 158, 162, 169, 170
Wood, 72–5, 104, 167
Woodwind instruments, 16

X-15, 18, 157

Yachts, 96–7